45

Progress in Mathematics

Volume 238

Series Editors

Hyman Bass

Joseph Oesterlé

Alan Weinstein

Harmonic Analysis, Signal Processing, and Complexity

Festschrift in Honor of the 60th Birthday of Carlos A. Berenstein

Irene Sabadini
Daniele C. Struppa
David F. Walnut
Editors

Birkhäuser
Boston • Basel • Berlin

Irene Sabadini
Politecnico di Milano
Dipartimento di Matematica
I-20133 Milano
Italy

Daniele C. Struppa and David F. Walnut
George Mason University
Department of Mathematical Sciences
Fairfax, VA 22030
U.S.A.

AMS Subject Classifications (2000): Primary: 05C05, 05C50, 31C20, 32A26, 32A50, 35C15, 35N05, 35R30, 42A85, 42B10, 42B35, 43A85, 44A12, 46F12, 46F15, 65R30, 92C55, 94A12

Library of Congress Cataloging-in-Publication Data
Harmonic analysis, signal processing, and complexity : festschrift in honor of the 60th
    birthday of Carlos A. Berenstein / Irene Sabadini, Daniele C. Struppa, David F. Walnut, editors.
        p. cm. – (Progress in mathematics ; v. 238)
    Includes bibliographical references.
    ISBN 0-8176-4358-3 (alk. paper)
        1. Harmonic analysis–Congresses. 2. Signal processing–Congresses. 3. Computational
    complexity–Congresses. I. Berenstein, Carlos A. II. Sabadini, Irene (Irene Maria), 1965-
    III. Struppa, Daniele Carlo, 1955- IV. Walnut, David F. V. Progress in mathematics (Boston, Mass.);
    v. 238.

QA403.H236 2005
515′.2433-dc22   2005045229

ISBN-10 0-8176-4358-3   e-IBSN 0-8176-4416-4   Printed on acid-free paper.
ISBN-13 978-0-8176-4358-4

©2005 Birkhäuser Boston   *Birkhäuser* ®

Printed in the United States of America.   (JLS/MP)

9 8 7 6 5 4 3 2 1   SPIN 11008651

www.birkhauser.com

*Carlos A. Berenstein, sitting in front of Leonhard Euler's tomb
in St. Petersburg, Russia.*

Contents

Preface

This volume consists of invited articles dedicated to Carlos Berenstein and the mathematics and mathematicians he inspired. A conference in Berenstein's honor was held on the occasion of his sixtieth birthday at George Mason University in May 2004.

We editors have been his students, collaborators, and friends for many years and as such will try in this preface to convey a very personal sense of the many dimensions of Carlos' work.

Those familiar with Berenstein's work will find that much has been omitted in this volume. No doubt, many other aspects of his mathematics could have been included.

We view this, however, as our personal tribute to a mathematician who has given so much to us and to the community, and we have tried to reflect the influence and impact of Carlos Berenstein's work in the articles here.

A quick glance at Berenstein's list of papers shows how difficult it is to provide a comprehensive analysis of the impact of his work, which encompasses pure harmonic analysis as well as applied topics such as optical deconvolution. His interests range from integral geometry on trees to microlocal analysis of theta functions, and yet his papers and books are connected by a series of ideas and problems that make them, taken as a whole, an important research program.

His first paper, published in CRAS in 1966, deals with interpolation of operators. In a different sense of interpolation, it is Carlos Berenstein's systematic use (with B. A. Taylor) of an old interpolation formula of Jacobi that forms the foundation of much of his more recent work. Similarly, some early theoretical papers on the three-squares theorem have given rise to an original and powerful theory of deconvolution. There are more examples: his work on harmonic analysis on discrete structures, and his contributions to mathematical tomography to name but two; but we wish to make a more general point here. In every problem Carlos attacked, and in every paper he wrote, we see the sign of his characteristic approach to mathematics: deep respect and understanding of the classical treatment of a subject, together with an ability and willingness to introduce the necessary innovation.

We mentioned the Jacobi interpolation formula. In Carlos' work with Taylor, that formula is transformed from an interesting historical curiosity into a fully developed and powerful tool to attack complex problems of interpolation in spaces of holomorphic functions with growth conditions. The Jacobi interpolation formula becomes the key instrument to prove a far-reaching generalization of Ehrenpreis' Fundamental Principle to convolution equations. Along similar lines, in joint work with with A. Yger, R. Gay, and A. Vidras and following earlier work of I. M. Gelfand, Jacobi's formula is extended in a systematic way to new settings and its counterpart in microlocal analysis is developed. The result is a new and powerful theory of residues.

A third example of Carlos' interest in connecting classical problems and ideas with modern techniques and innovation has to do with his deep work on the Nullstellensatz and Bezout identities. The problems were well known and the subject is classical, yet the clever and technically arduous approach of Carlos and his collaborators is innovative and extremely effective.

The final and most striking example we wish to mention is given by the two complex analysis textbooks he coauthored with Roger Gay. Those books are a pleasure to read, not only for their clarity and innovative approach, but also because they reopen very classical questions such as those related to issues of overconvergence, and address them in a totally novel way. Those books, we believe, could and should be studied by younger generations of mathematicians. The ideas presented are likely to offer much food for thought and foster additional advances.

One thing we note by perusing Berenstein's list of publications is his increasing interest in far-reaching applications of mathematics. Signal processing, tomography, and deconvolution have become staples of his publications and effectively build a bridge between mathematics and its manifold applications. And yet if we look at the development of Berenstein's mathematics we see that these interests are evident even in his early work on interpolation, on Ehrenpreis' Fundamental Principle, and on the arithmetical questions needed to understand the behavior of exponential polynomials. Carlos' old Springer Lecture Notes (coauthored with M. Dostal) include some ideas on exponential sums and polynomials which later flourished into a treatment of specific deconvolution problems. Signal processing is one of his later interests, yet we see traces of what would eventually become engineering applications in many of his early papers. Even his interest in the Nullstellensatz and Bezout identities grew in tandem with his interest in questions related to systems control. Those of us who studied and worked with him were taught that there is no separation between pure and applied mathematics; they merge and enrich each other through constant osmosis.

But Carlos' contribution to mathematics is not limited to his scientific output. Equally important is the incredibly strong and steady support he gives to his students, friends, and colleagues. Carlos' generosity has led him to direct, so far, sixteen dissertations (many of his students participated in this volume as a way of showing him their gratitude and respect) and his published work involves more than fifty coauthors! Carlos is quick to acknowledge the role of his coworkers, even when that role has been minor, or limited to a few conversations. His policy is that if he talks about a problem with a colleague then that colleague has earned a place on the paper. At the same time he easily and willingly lends his ideas to other mathematicians without asking for any formal acknowledgment.

Carlos is a generous man and is constitutionally incapable of making a negative remark about anybody. As few other mathematicians can, he finds something good and valuable in any remark. His understanding of mathematics is such that even a trivial comment may be transformed in his mind into an exciting new way of thinking about an old problem. It has happened to at least one of us (dcs): a rather trivial comment on comparison theorems for convolution operators has become the basis for some really intriguing work on Dirichlet series, work which later became the starting point of further analysis by mathematicians in several parts of the world.

Many of us have one anecdote or another in which Carlos' generosity of ideas and of support come through with rare force. We will offer a few such examples, though we are sure that each of the contributors to the volume could add their own. One of us (dcs) still remembers a time early in his career when he was collaborating with Carlos on some problems, which another young mathematician was pursuing simultaneously in a different country. He remembers vividly how Carlos suggested not trying to occupy the entire territory but leaving some room for future collaborations. The desired collaboration eventually took place, and he learned the importance of thinking of mathematics as a collaborative effort among a community that shares a passion, an interest, and a sense of healthy, good-natured competition.

Another of us (dw) remembers a year spent with Carlos as a colleague in the Mathematical Sciences Department at George Mason University. In weekly meetings, Carlos would be a fountain of ideas for work on problems in tomography, deconvolution, sampling, signal processing, wavelets, and more. While only a few of these wonderful ideas were brought to fruition in the form of joint publications (through no fault of Carlos, to be sure) they did form the basis in later years of Masters and Doctoral dissertations as well as sufficient spinoff ideas to sustain an interesting and productive research program for many years. It was also encouraging to experience such a first-rate mind at a formative stage in one's early career.

As we join with many friends to celebrate Carlos' sixtieth birthday, we wish him (and us) many more years of successful and significant work.

Irene Sabadini
Daniele C. Struppa
David F. Walnut
Fairfax, January 2005

Harmonic Analysis,
Signal Processing,
and Complexity

Some Novel Aspects of the Cauchy Problem

Leon Ehrenpreis

Department of Mathematics
Temple University
Philadelphia, PA 19122
USA
leonzeta@aol.com

To Carlos.
It has been an honor to have you as my student.

Summary. We discuss two subtle Cauchy problems:

1. What happens to the Cauchy–Kowalewsky theorem when the initial data is not analytic? We show that there is a solution to the asymptotic Cauchy problem, meaning that Pf is not zero, but approaches 0 rapidly as we approach the Cauchy surface. The rapidity depends on the regularity of the initial data.
2. We introduce a new type of Cauchy problem. In particular, we show how it applies to the parametric Radon transform and gives a new insight into the sufficiency of the John equations.

1 Introduction

Let $\overrightarrow{P}(x, y; D) = [P^1(x, y; D), \ldots, P^r(x, y; D)]$ be a system of linear differential operators. We write $\overrightarrow{P} f = 0$ for $P^j f = 0$ for all j. For the Cauchy Problem (CP) we are given a Cauchy Surface (CS) S (dim $S = m$) and partial differential operators $q_1, \ldots, q_s$ which are not tangent to S. The CP is the study of the map

$$\mathcal{C} : f \to (q_1 f_{|S}, \ldots, q_s f_{|S}; \overrightarrow{P} f). \tag{1}$$

$\mathcal{C}$ maps functions of x, y into s tuples of functions on S and r tuples of functions in $\mathbb{R}^n$. We assume S is "nice" so we have a coordinate system (x, y) in the ambient space with S defined by $\{ y = 0 \}$, dim$\{x\} = n$, dim$\{y\} = m$. (In some cases we must add restrictions of derivatives of f to subsets of S. This leads to complications which can be easily overcome; they do not occur in our examples.)

For the classical CP we have $r = 1$, S is a hypersurface and q_j are powers of a nontangential derivative. The theory for $r > 1$ was begun by Cartan and Kähler; for constant coefficient operators it is exposed in detail in [E1]. We shall deal only with the kernel of $\overrightarrow{P}$, i.e., $\overrightarrow{P} f = 0$.

For $r = 1$ the classical theory has two directions:

(a) *Cauchy–Kowalevsky theorem.* The coefficients of $\overrightarrow{P} = P$ are assumed to be holomorphic near the origin and S is a noncharacteristic hypersurface for P. Then C is an isomorphism in the setting of locally holomorphic functions.

(b) *Hyperbolic equations.* S is a space-like hypersurface for P. Then "holomorphicity" in the Cauchy–Kowalevsky theorem can be replaced by C^∞ or another class of smooth functions.

Our goal is to present some interesting modifications of (a) and (b).

Our second ramification of the CP has its inception in the parametric Radon transform [E2, Chapter 6].

Let f be a function on $\mathbb{R}^n$. We define the parametric line Radon transform of f to be

$$Jf(a, b) = \int f(a\lambda + b)d\lambda. \tag{2}$$

Here λ is a real parameter and a, b are points in $\mathbb{R}^n$, so Jf describes the integrals of f over all lines in $\mathbb{R}^n$. Note that Jf satisfies certain equations, called the *John equations*. These are

$$\frac{\partial^2}{\partial a_i \partial b_j} Jf = \frac{\partial^2}{\partial a_j \partial b_i} Jf, \quad i, j = 1, \ldots, n \tag{3}$$

provided f is smooth, which we assume. Equation (3) holds even if the measure $d\lambda$ is replaced by an arbitrary measure $d\mu(\lambda)$. The Haar measure property of $d\lambda$ can be expressed by

$$Jf(a, b + a\lambda^0) = Jf(a, b) \tag{4}$$

or, in infinitesimal form,

$$(a \cdot \nabla_b)Jf = 0. \tag{5}$$

We would like to set up a CP for the John equations and, somehow, construct f from Jf and its Cauchy Data (CD). To this end note that

$$J_\lambda(a, b) = f(a\lambda + b) \tag{6}$$

satisfies the first order system

$$\left(\frac{\partial}{\partial a_i} - \lambda \frac{\partial}{\partial b_i} \right) J_\lambda f = 0, \quad i = 1, 2, \ldots, n. \tag{7}$$

Moreover $f = f(b)$ is the CD for this system on the Cauchy Surface (CS) $a = 0$. We cannot set $a = 0$ in (2) because the integral does not exist. (It is shown in [E2, Chapter 4] that Jf is locally integrable in a, b and satisfies the John equations in the distribution sense for suitable f provided that $n \geq 3$.)

The expression (2) means that Jf is an integral of solutions of the first order equations (7). Such "reduction" of solutions of high order equations to integrals of solutions of first order equations is the essence of twistor theory. The simplest illustration of twistor theory is the expression of an harmonic function on $\mathbb{R}^2$ (solution of second order equation) as the sum of an holomorphic function and an antiholomorphic function. Whittaker and Bateman extended this idea to harmonic functions in $\mathbb{R}^3$ and $\mathbb{R}^4$ (see [E2, Section 1.4]).

We call an equation (system of equations) $Pf = 0$ an *enveloping equation* of a family of equations $^k P f_k = 0$ when each f_k is also a solution of $P f_k = 0$ and all solutions of $Pf = 0$ are limits of linear combinations of such f_k. $^k P$ are called the twistor operators for P. Thus Δ is the enveloping operator of the first order operators that appear in the work of Bateman, Whittaker and Penrose. We shall show that J is the enveloping operator of the J_λ.

2 The asymptotic Cauchy–Kowalevsky theorem

Suppose that S is analytic and the coefficients of P are holomorphic but the Cauchy data (CD), which is the set of s functions $g_1, \ldots, g_s$ on S which potentially represent $\{q_j f_{|S}\}$, are not holomorphic. What can we say? To understand what is happening, let us contrast the operators ($r = 1, n = m = 1, S = \{y = 0\}$):

$$^1 P \equiv \frac{\partial}{\partial y} - \frac{\partial}{\partial x}, \qquad ^2 P \equiv \frac{\partial}{\partial y} + i \frac{\partial}{\partial x} = \frac{\partial}{\partial \bar{z}}, \qquad (8)$$

$$q_1 = \text{identity} .$$

$^1 P$ is hyperbolic and $^2 P$ is elliptic. If $f \in C^\infty$, then $^1 P f = 0$ imposes no regularity on g except the obvious condition that $g \in C^\infty$. On the other hand if $^2 P = 0$ in a neighborhood of S, then g is analytic. If $^2 P = 0$ in a one-sided neighborhood of S, say $y > 0$, and f has suitable boundary values on S, then the analytic wave front set of g omits the positive axis. Conversely, if the analytic wave front set of g omits the positive axis, then there exists a solution of the CP for $\partial / \partial \bar{z}$ (locally) in $y > 0$ with CD $(f) = g$.

Let us pass to the situation in which g is not analytic. The classes of g we consider are Gevrey ℓ classes G_ℓ. This means that

$$|g^{(m)}(x)| \le C^{m+1} (m\ell)!. \qquad (9)$$

We shall also consider Gevrey front sets. These are studied in [E2, Chapter 5]. We shall return to them later.

Let us search for a formal solution of the CP

$$f(y, t) = \sum f_m y^m, \qquad (10)$$

where f_m are functions on S. Our equation asserts

$$(m + 1) f_{m+1} + i f'_m = 0. \qquad (11)$$

By iteration

$$f_m = \frac{1}{m!} i^m g^{(m)}. \tag{12}$$

This means that

$$\left| \frac{\partial^m}{\partial y^m} f(0, x) \right| = |m! f_m| = |i^m g^{(m)}| \le C^{m+1}(m\ell)!. \tag{13}$$

Of course this is not sufficient to make the series (10) converge. But there is a theorem due to the author and to B. Mitjagin (see [E1, Chapter XIII]) which can be modified to show that there is a function $F(x, y) \in G_\ell$ such that

$$\frac{\partial^m}{\partial y^m} F(0, x) = \frac{\partial^m}{\partial y^m} f(0, x). \tag{14}$$

(This is a quantitative sharpening of the classical theorem of E. Borel which asserts that for any sequence $\{a_m\}$ of numbers there is a C^∞ function $A(y)$ with $A^{(m)}(0) = a_m$.) Of course F is *not* a solution of $\bar\partial F = 0$ but F is an *asymptotic solution* meaning

$$H(y, x) \equiv \left(\frac{\partial}{\partial y} + i \frac{\partial}{\partial x} \right) F(y, x) \to 0 \quad \text{rapidly as } t \to 0. \tag{15}$$

How rapidly?

Note that, by construction, $H \in G_2$ and its formal power series is identically 0 on $\{y = 0\}$. This means that F is a formal solution of $\partial/\partial\bar z$ on $y = 0$. Since $H \in G_2$, we can estimate how rapidly $H \to 0$ as $y \to 0$. We could use the mean value theorem or, more simply, write

$$|H(y, x)| = \left| \int_0^y dy_{m-1} \int_0^{y_{m-1}} dy_{m-1} \cdots \int_0^{y_1} \frac{\partial^m}{\partial y^m} H(y_0, x) dy_0 \right| \tag{16}$$

$$\le \frac{C^{m+1}|y|^m(\ell m)!}{m!}$$

uniformly on compact sets. The best bound for H comes from choosing m optimally. Except for a change of the constant we can write

$$|y|^m \frac{(\ell m)!}{m!} \sim |y|^m m^{(\ell-1)m}. \tag{17}$$

By differentiation the minimum occurs at

$$m \sim |y|^{-(\ell-1)^{-1}} e^{-1}. \tag{18}$$

The minimum gives

$$|H(y, x)| \le C e^{-c|y|^{-1/(\ell-1)}}. \tag{19}$$

In particular, for $\ell = 2$ the minimum gives

$$|H(y, x)| \le C e^{-c'/|y|}. \tag{20}$$

When H satisfies (19) we call F an *asymptotic solution of order* $e^{-|y|^{-1/(\ell-1)}}$. In fact, our derivation of (14) would apply equally to ^1P. But, of course, we can do better for ^1P. Since ^1P is hyperbolic, if $g \in C^\infty$ then there exists a true solution of $^1Pf = 0$, $f \in C^\infty$ with $CD(f) = g$. For such an f, (19) would hold for all l (since, in fact, $H = 0$) so there is no possibility of reversing (19) to deriving regularity of g.

However, we shall derive a converse for real elliptic systems. There may be a converse for certain hypoelliptic systems but we have not pursued this possibility.

3 Gevrey classes lead to asymptotic solutions

Let us begin with the case $r = 1$. We assume that y is a single variable and $x = (x_1, \ldots, x_n)$. Let S be a real analytic hypersurface which is not characteristic for P. By changing variables we may assume that

$$P \equiv \frac{\partial^s}{\partial y^s} + P_1\left(x, y, \frac{\partial}{\partial x}\right)\frac{\partial^{s-1}}{\partial y^{s-1}} + \cdots + P_s\left(x, y, \frac{\partial}{\partial x}\right). \tag{21}$$

The coefficients of the P_j are in G_ℓ. The noncharacteristic nature of S means that order $P_j \leq j$. We fix the CP by using the operators

$$q_j = \frac{\partial^j}{\partial y^j}, \quad j = 0, 1, \ldots, s - 1. \tag{22}$$

The CD is $\{g_j\}$. We express the formal power series of f as in (10). From (21) we deduce

$$s! f_s = -\sum_0^{s-1} j! P_{s-j}\left(x, 0, \frac{\partial}{\partial x}\right) g_j. \tag{23}$$

This expresses f_s in terms of the CD. In a similar way, equation (21) shows how to write f_p for $p \geq s$ in terms of $f_{p'}$ for $p' < p$, and hence, by iteration, in terms of the CD $\{g_j\}$. Precisely,

$$p(p-1)\cdots(p-s+1)f_p = -\sum_0^{s-1}(p-j_1)\cdots(p-s+1)P_{j_1}f_{p-j_1} \tag{24}$$

and noting that

$$\frac{(p-j_1)\cdots(p-s+1)}{p(p-1)\cdots(p-s+1)} = \frac{(p-j_1)!}{(p-s)!}\frac{(p-s)!}{p!},$$

we deduce, by setting $j_0 = 0$,

$$f_p = -\sum \frac{(p-j_1)!}{(p-j_0)!}P_{j_1}f_{p-j_1} = \sum \pm \frac{(p-j_1)!}{(p-j_0)!}\frac{(p-j_1-j_2)!}{(p-j_1)!}P_{j_1}P_{j_2}f_{p-j_1-j_2} \tag{25}$$

$$= \cdots = \sum \pm \frac{(p-j_1-j_2-\cdots-j_n)!}{(p-j_0)!}P_{j_1}P_{j_2}\cdots P_{j_n}f_{p-j_1-\cdots-j_n}.$$

The sum is over all $j_1, \ldots, j_n$ such that

$$0 \le p - \sum j_i < s. \tag{26}$$

Once (26) is satisfied we can replace $f_{p-j_1-\cdots-j_n}$ by $g_{p-j_1-\cdots-j_n}$ times the insignificant constant $[(p - j_1 - \cdots - j_n)!]^{-1}$.

Let us take stock of bounds. $(p - j_1 - \cdots - j_n)!/p!$ is clearly of the order $(p!)^{-1}$. The number of terms in the sum is certainly bounded by s^p since $n \le p$ and each j_i is bounded by $s - 1$. Since factors of the form c^p are insignificant for bounds in Gevrey classes, we can ignore them. It remains to estimate individual terms of the form

$$P_{j_1} \ldots P_{j_n} g_v$$

when $0 \le v = p - \sum j_i < s$. Note that $P_{j_1} \ldots P_{j_n}$ is a differential operator of order $\le \sum j_i \le p$. It seems reasonable that the effect on any $g_k \in G_\ell$ should be bounded by $C^p(\ell p)!$. Indeed, this is the case as we now show. It is instructive to understand the case $n = 1$, $m = 1$ as the general situation follows the same lines. P_j is of the form

$$P_j = \sum_0^j \alpha_{jk} \frac{\partial^k}{\partial x^k}, \tag{27}$$

where $\alpha_{jk} \in G_\ell$. Let us compute the "norm" of P_j on G_ℓ meaning let us find the bound for

$$\alpha_{jk} \frac{\partial^k}{\partial x^k} g \tag{28}$$

in terms of the G_ℓ bounds on $g \in G_\ell$. To accomplish this we first find bounds for the "norm" of $\partial^k/\partial x^k$ and then for the "norm" of multiplication by α_{jk}. We have

$$\left| \frac{\partial^{t+k}}{\partial y^{t+k}} f \right| \le C^{t+k}[\ell(t+k)]! \le C_1^{t+k}(t+k)^{\ell(t+k)} \tag{29}$$

$$= C_1^{t+k} \left(\frac{t+k}{t} \right)^{\ell t} \left(\frac{t+k}{t} \right)^{\ell k} t^{\ell(t+k)} \le C_2^{t+k} e^{k\ell t} 2^{\ell k} [\ell(t+k)]!$$

as long as $t \ge k$, which we may assume since $k \le s$.

We have shown that the "norm" of f, which we can think of as the constant C in the first line of (29), is multiplied by a fixed constant when f is differentiated k times. We now present a similar calculation for the operator of multiplication by a function $\alpha \in G_\ell$. We have

$$\left| \frac{\partial^t}{\partial y^t} \alpha f \right| \le 2^t \max \alpha^{(k)} f^{(t-k)} \le C_3^t (\ell k)! [\ell(t-k)]!. \tag{30}$$

Now $(\ell k)![\ell(t - k)]!/(\ell t)!$ is a reciprocal binomial coefficient and so is bounded by 1. Since the number of differentiations and multiplications in each P_j is bounded by a fixed constant times order P_j, we deduce that the "norm" of $P_{j_1} \ldots P_{j_n}$ is bounded by C^p. Hence, by (25) with $\{j_i\}$ as in (21)

$$|f_p| \leq C_4^p \frac{(\ell p)!}{p!} \leq C_5^p[(\ell - 1)p]!. \tag{31}$$

The extension of this argument to $n > 1$ presents no difficulties.

How do we formulate the CP problem when $r > 1$ in order to derive the same result? In the case $r = 1$ we used (see (1)) $q_j = \partial^j/\partial y^j$, $j = 0, \ldots, s - 1$. These operators had the following important property:

For any $a = (a_1, \ldots, a_m)$, we can write

$$\frac{\partial^a}{\partial y^a} = \sum \mathcal{U}_j^a q_j \text{ mod } \overrightarrow{P}, \tag{32}$$

where

$$\text{order} \left[\mathcal{U}_j^a = \mathcal{U}_j^a \left(x, y, \frac{\partial}{\partial x} \right) \right] \leq |a| - \text{order } q_j. \tag{33}$$

"mod $\overrightarrow{P}$" means modulo the left module generated by the P_j in the ring of differential operators ($\mathcal{D}$-module). Actually, we need (32) only for

$$|a| \leq \text{max order } q_j. \tag{34}$$

We now define the general noncharacteristic CP.

Noncharacteristic CP. (32), (33) are valid for all a satisfying (34). The q_j are assumed to be operators in y only. Then (32), (33) imply that for any j, k we can write

$$\frac{\partial}{\partial y_k} q_j = \sum V_{jk\ell} q_\ell, \tag{35}$$

where $V_{jk\ell} = V_{jk\ell}(x, y, \frac{\partial}{\partial x})$ satisfies

$$\text{order } V_{jk\ell} \leq \text{order } q_k + 1. \tag{36}$$

With this setup we can mimic the argument leading to (31). Then we can apply the integration idea of (16). This leads to the following.

Theorem 1. Let $\overrightarrow{P}$ be a system of operators with Cauchy Problem satisfying (32), (33), (34) and coefficients in G_ℓ. Then the CP with data in G_ℓ has an asymptotic solution of order

$$e^{-|y|^{-1/(\ell-1)}}. \tag{37}$$

We now prove the converse. As we noted in the Introduction, the converse is "conditional."

We shall restrict our consideration to $r = 1$. Let f be a smooth function of x, y. We suppose f is defined for all x although we could work locally using various techniques (see, e.g., [E2, Chapter 5]).

Call $\overrightarrow{f}_y$ the CD of f on the hyperplane defined by fixing y. Thus

$$\overrightarrow{f}_y(x) = \left(f(x, y), \frac{\partial}{\partial y} f(x, y), \ldots, \frac{\partial^{s-1}}{\partial y^{s-1}} f(x, y) \right). \tag{38}$$

It is easier to express the equation $Pf = 0$ as a system

$$\frac{\partial \overrightarrow{f}}{\partial y} = \boxed{\mathrm{P}} \overrightarrow{f}. \tag{39}$$

$\boxed{\mathrm{P}}$ is a matrix of operators in x with coefficients depending on (x, y). Let us start with constant coefficient P. Then $\boxed{\widehat{\mathrm{P}}}$, the Fourier transform of $\boxed{\mathrm{P}}$, is a matrix of polynomials in $\hat{x}$. The characteristic roots $\lambda_j(\hat{x})$ of $\boxed{\mathrm{P}}$ are (up to a factor i) the roots of $\hat{P}(\hat{x}, \hat{y}) = 0$. ($\hat{P}$ is thought as a polynomial in $\hat{y}$ with coefficients which are polynomials in $\hat{x}$.) We assume for the present that these roots are generically distinct. (The factor i comes from the fact that in (39) we do not take Fourier transform in y as we do in $\hat{P}(\hat{x}, \hat{y})$.)

We now make our crucial *assumption*. P is elliptic.

Our assumption means that the roots satisfy

$$|\operatorname{Re} \lambda_j(\hat{x})| \geq C|\hat{x}| \tag{40}$$

for $\hat{x}$ large. We now suppose that, in a suitable sense,

$$\left\| \frac{\partial \overrightarrow{f}}{\partial y} - \boxed{\mathrm{P}} \overrightarrow{f} \right\| \leq \alpha(y), \tag{41}$$

where, in accordance with (37), we shall usually set $\alpha(y) = \exp(-|y|^{-1/(\ell-1)})$. We want to show that this forces regularity on the CD which is $\overrightarrow{f}(0)$. If f is small at infinity in x we can take its Fourier transform in the usual sense. If not, we can use some form of "cut-offs" as described in [E2, Chapter 5] using the nonlinear Fourier transform. We shall proceed formally; a justification can be given using the methods of [E2, Chapter 9].

Taking Fourier transform in x, we derive

$$\left\| \frac{\partial \overrightarrow{\hat{f}}}{\partial y} - \boxed{\widehat{\mathrm{P}}} \overrightarrow{\hat{f}} \right\| \leq \alpha(y). \tag{42}$$

Fix $\hat{x}$ and express $\overrightarrow{\hat{f}}(\hat{x})$ in terms of the eigenvectors of $\boxed{\widehat{\mathrm{P}}}(\hat{x})$. Thus

$$\overrightarrow{\hat{f}}_y(\hat{x}) = \sum a_j(y) \hat{\mu}_j(\hat{x}), \tag{43}$$

where $\hat{\mu}_j(\hat{x})$ are the eigenvectors, with eigenvalues $\lambda_j(\hat{x})$. (Of course, a_j depend on $\hat{x}$.) We can normalize the $\hat{\mu}_j$ by requiring that $\|\hat{\mu}_j(\hat{x})\| = 1$ for each j. This leaves a further normalization which is of no consequence. The complement of the set of $\hat{x}$

where $\hat{\mu}_j(\hat{x}) = 0$ for some j is sufficient. As we have explained in [E2, Chapter 5], we may assume that the eigenvalues $\lambda_j(\hat{x})$ are distinct since, if P has no multiple factors, the set of $\hat{x}$ for which the eigenvalues are distinct is sufficient. For each j, we integrate the inequality (42). We write

$$\frac{\partial}{\partial y} - \lambda_j(\hat{x}) = e^{\lambda_j y}\frac{\partial}{\partial y}e^{-\lambda_j y} \tag{44}$$

so that, on integrating over $[0, y]$,

$$e^{-\lambda_j y}a_j(y) = a_j(0) + O\left[\int_0^y e^{-\lambda_j y}\alpha(t)dt\right]. \tag{45}$$

We want to show that $a_j(0)$ are small as functions of $\hat{x}$. We have bounds on $a_j(y)$ which come purely from the regularity of f, that is, such bounds do not depend on the equation. But we can obtain better bounds on $a_j(0)$, as functions of $\hat{x}$ because of the factors $\exp(-\lambda_j y)$ and because of the fact that $\alpha(t)$ is small.

We noted in (40) that $\mathrm{Re}\,\lambda_j(\hat{x})$ is large. In order to use this fact we choose $y > 0$ or $y < 0$ depending on whether $\mathrm{sgn}\,\mathrm{Re}\,\lambda_j(\hat{x})$ is $+$ or $-$. Suppose that $\mathrm{Re}\,\lambda_j(\hat{x}) > 0$. Then we choose $y > 0$. Making y large forces the term $\exp(-\lambda_j y)a_j(y)$ to be small, but it loses the strong decrease of $\alpha(t)$ as $t \to 0$ in the integral in (45). Thus we seek an optimal choice for y in (45). Write

$$\alpha(t) = e^{\beta(t)} \tag{46}$$

so $\beta(t) \to \infty$ as $t \to 0$. We choose y at the maximal point of the integrand.

$$\max_t[-\lambda_j t + \beta(t)] = \tilde{\beta}(\lambda_j) \tag{47}$$

is a form of the Legendre transform (Young conjugate) of β.

In relation to Theorem 1, $\beta(t) = t^a$ for some $a < 0$, so, from (47) $\lambda_j = ay^{a-1}$ or $\lambda_j y = ay^a = a\beta(y)$ at the maximal point, which we choose for y. This gives

$$y = c\lambda_j^{1/a-1}. \tag{48}$$

Combining this with (45) leads to

$$a_j(0) \leq c_1 \exp(-c\lambda_j^{a/a-1}) + c\lambda_j^{1/a-1}\exp[c_2(1-a)\lambda_j^{a/a-1}]. \tag{49}$$

The constants c, $c_2(1-a)$ which multiply $\lambda_j^{a/a-1}$ are not significant. Thus we have the essential bound

$$a_j(0) \leq c_3 \exp(-c_4\lambda_j^{a/a-1}). \tag{50}$$

To coordinate this with Theorem 1, we set $a = -1/(\ell - 1)$. Thus

$$a_j(0) \leq c_3 \exp(-c_4\lambda_j^{1/\ell}). \tag{51}$$

We now apply the ellipticity assumption (40). Since $\mathrm{Re}\,\lambda_j > 0$ we deduce

$$a_j(0) \le c_5 \exp(-c_4 |\hat{x}|^{1/\ell}). \tag{52}$$

When Re $\lambda_j < 0$ we integrate over $(0, -y)$ and argue as before. We conclude the following.

Theorem 2. *The converse of Theorem 1 is valid for elliptic equations with constant coefficients.*

What happens when the coefficients are variable? It appears that if the coefficients do not depend on t, then we can modify our method by diagonalizing the matrix $\boxed{P}$ using Fourier integral operators. We could then proceed in an analogous (but more complicated) fashion.

When the coefficients depend on y also, it seems that further ideas are needed. We leave this to the reader.

4 The twistor Cauchy Problem

Let us return to the notation in the Introduction, in particular to the John equation (3). By the Fundamental Principle (see [E2, Section 1.4]), we expect that solutions of (3) are spanned by exponential solutions. (The Fundamental Principle per se does not apply in this case because Jf is not in an AU space. Nevertheless, in [E2, Chapter 6] we clarify the necessary modifications.) These remarks mean that solutions F of

$$\left(\frac{\partial^2}{\partial a_i \partial b_j} - \frac{\partial^2}{\partial a_j \partial b_i} \right) F = 0 \quad (i.j = 1, 2, \ldots, n) \tag{53}$$

can be represented as Fourier transforms on the algebraic variety V related to (53), namely,

$$\hat{a}_i \hat{b}_j = \hat{a}_j \hat{b}_i. \tag{54}$$

Equation (54) means that $\hat{a} \parallel \hat{b}$. For later purpose we can rephrase this as follows: the $n \times 2$ matrix formed by $\hat{a}, \hat{b}$ (thought of as column vectors) has rank 1.

Any $\hat{a} \ne 0$ defines the line

$$V_{\hat{a}} = \{(\hat{a}, \lambda \hat{a})\}_\lambda. \tag{55}$$

However, we are interested in a different decomposition of V. The variety V_λ associated to the system J_λ (see (7)) is given by (λ fixed)

$$V_\lambda : \quad \hat{a} = \lambda \hat{b}. \tag{56}$$

Instead of thinking of V as fibred by the lines $V_{\hat{a}}$ which depend on $\hat{a}$, we regard V as fibred by the V_λ which are of dimension n. According to the ideas of [E1, Chapter IX], the Fourier transform $\widehat{C}_\lambda^t$ of the adjoint of the Cauchy map C_λ is the extension of functions $\hat{h}(\hat{b})$ to V_λ given by

$$(\widehat{C}_\lambda^t \hat{h})(\hat{a}, \hat{b}) = \hat{h}(\hat{b}) \quad \text{on } V_\lambda. \tag{57}$$

This means that $\widehat{C'_\lambda}\hat{h}(\lambda\hat{b},\hat{b}) = \hat{h}(\hat{b})$. By duality, (57) becomes

$$(\widehat{C_\lambda}\hat{\alpha}_\lambda)(\hat{b}) = \hat{\alpha}_\lambda(\lambda\hat{a},\hat{b}). \tag{58}$$

In (58) $\hat{\alpha}_\lambda$ is the Fourier transform of a solution of J_λ; it is an "arbitrary" function on V_λ. We say "arbitrary" because it is only restricted by growth and regularity. (This is the Fundamental Principle for V_λ.)

Note that $\cup V_\lambda$ is "almost" all of V. The only part of V missing from $\cup V_\lambda$ is the $\hat{a}$-axis, except for the origin. If we impose a suitable regularity on functions $\hat{\alpha}$ (more exactly, measures) on V then we can write

$$\hat{\alpha} = \int \hat{\alpha}_\lambda d\mu(\lambda), \tag{59}$$

where $\hat{\alpha}_\lambda$ is a measure on V_λ.

Let us return to the situation described by the Radon transform. Call α the parametric Radon transform of f; in the notation of (2), $\alpha = Jf$. α is a solution of the John equation (3) so α is the Fourier transform of a measure $\hat{\alpha}$ on V. Moreover, α satisfies the *invariance equation* (5). By Fourier transform

$$(\hat{b} \cdot \nabla_{\hat{a}})\hat{\alpha} = 0 \tag{60}$$

that is, for any λ^0,

$$\hat{\alpha}(\hat{a} + \hat{b}\lambda^0, \hat{b}) = \hat{\alpha}(\hat{a},\hat{b}). \tag{61}$$

We want to determine f so that $\alpha = Jf$. If such an f existed, then, by the above, there would exist functions α_λ which satisfy $J_\lambda\alpha_\lambda = 0$, and all α_λ have the same CD for J_λ. Moreover,

$$\alpha = \int \alpha_\lambda d\lambda.$$

By Fourier transform this means that we can decompose $\hat{\alpha}$ into $\hat{\alpha}_\lambda$ with support $\hat{\alpha}_\lambda$ on V_λ and $\widehat{C_\lambda}\hat{\alpha}_\lambda$ independent of λ; by the above this means $\hat{\alpha}$ is independent of λ. Since V_λ is defined by $\{\hat{a} = \lambda\hat{b}\}$, equations (61) and (54) show that $\hat{\alpha}(\hat{a},\hat{b})$ depends only on $\hat{b}$ so the desired decomposition of $\hat{\alpha}$ can be effected.

We can go further in this direction. Instead of assuming that all $\widehat{C_\lambda}\hat{\alpha}_\lambda$ are equal, suppose there is a function $\mathscr{A}(\lambda)$, with $\mathscr{A}(1) = 1$, for which

$$\widehat{C}\hat{\alpha}_\lambda = \mathscr{A}(\lambda)\widehat{C}\hat{\alpha}_1. \tag{62}$$

This means that all α_λ have the same CP on $a = 0$ up to a constant multiple. We can reinterpret this by modifying (2) to

$$J_\mathscr{A} f(a,b) = \int f(a\lambda + b)\mathscr{A}(\lambda)d\lambda. \tag{63}$$

f is the Fourier transform of $\widehat{C}\hat{\alpha}_1$. We have shown how this "twistor CP" impinges on the parametric Radon transform and clarifies the sufficiency of the John equations.

We can reformulate equation (62) as follows. The variety V has a parametrization almost everywhere by $\hat{b}$, λ. Our condition on $\hat{\alpha}$ is: for any $(\lambda, \hat{b})$, if $\hat{a} = \lambda\hat{b}$ then

$$\hat{\alpha}(\hat{a}, \hat{b}) = \hat{\alpha}(\lambda\hat{b}, \hat{b}) = \hat{f}(\hat{b})\mathscr{A}(\lambda). \tag{64}$$

Of course, there is a serious technical problem to show that $\{(\lambda\hat{b}, \hat{b})\}$ is sufficient. This depends on the class of f considered. The problem is dealt with in [E2, Chapter 6].

We can go further. As noted in the Introduction, John's equation (3) is valid even when the measure $d\lambda$ in (2) is replaced by an arbitrary measure (even a suitable type of distribution) μ. Of course, the invariance equation (4) or (5) has to be modified. Call J^μ the resulting Radon transform. $J^\mu f$ is still an integral of $\{J_\lambda f\}$; only the measure is changed. Thus

$$\widehat{J^\mu f} = \int \widehat{J_\lambda f}\, d\mu(\lambda). \tag{65}$$

$J_\lambda f$ are independent of μ. Equation (65) shows that $\widehat{J^\mu f}$ is the product of $\widehat{J_1 f}$, which is a function of $\hat{b}$, with $\mu(\lambda)$ in the coordinates $(\hat{b}, \lambda)$, where $\hat{a} = \lambda\hat{b}$. (We are using the word "function" to include measure and distribution.)

Conversely, if α is a function of (a, b) which satisfies John's equation for which $\hat{\alpha}$ is a product of functions of $\hat{b}$ and of λ, then

$$\alpha = J^\mu f \tag{66}$$

for some μ, f. Since λ and b are variables on noncompact sets we have no way of normalizing f, μ so they are unique up to a constant multiple. We have shown the following.

Theorem 3. *A function $F(a, b)$ (with suitable restrictions on growth and smoothness) is of the form $J^\mu f$ for some suitable μ, f if and only if F satisfies the John equations and $\hat{F}$ (which has support in V) is the product of $\hat{f}(\hat{b})$ with $\mu(\lambda)$.*

Remark. The twistor method presented here appears to be simpler than the group convolution idea involved in [E2, *Theorems* 6.6 *and* 6.7].

These ideas also lead to the formulation of the *Twister–Watergate Problem* (TWP). The Watergate Problem (WP) is the analogue of the CP when the CS has dimension lower than the "natural" dimension of the operator $\overrightarrow{P}$. When $\overrightarrow{P}$ has constant coefficients, the natural dimension is the dimension of the associated algebraic variety. In particular, the natural dimension of the John equations is $n + 1$ where $n = \dim\{x\}$. As an example for the TWP we give data on the b-axis.

We can understand this TWP by Fourier transform. We have noted that those functions on V which are of interest to us are supported by $\{\hat{b}, \lambda\}$. Instead of picking a fixed function of λ as above, we now pick a basis for functions of λ, for example the Fourier basis $\{e^{i\lambda\hat{\lambda}}\}$. According to the ideas of [E2, Chapters 3 and 4] the WP involves the expansion of functions on V in terms of products $e^{i\lambda\hat{\lambda}}\hat{\alpha}(\hat{b})$. When $\hat{\lambda}$ is fixed there are the Fourier transforms of functions of the form

$$J^{\hat{\lambda}}f(a, b) = \int f(a\lambda + b)e^{i\lambda\hat{\lambda}}d\lambda.$$ (67)

We call this the TWP because we relate it to the CP by using the decomposition of "most of" V into a union of $V_\lambda = \{\hat{a} = \lambda\hat{b}\}$. This is necessary because $\{\lambda\}$ is not compact. (In the compact case, such as the WP for the wave equation with CS $= t$-axis, we do not need the "twistor breakup.")

We can go further in this direction. Let $\mathcal{U}$ be an algebraic variety which is fibred by subvarieties $\mathcal{U}_\lambda$. As an example, let us examine the WP for the wave equation in three dimensions from the twistor viewpoint. The analogue of J_λ is

$$J_\lambda(\Box)f(t, \lambda) = f(t\cos\lambda, t\sin\lambda, t),$$ (68)

which is clearly a solution of the wave equation. More precisely, $J_\lambda(\Box)f$ is the solution of the CP for

$$\begin{cases} \left(\frac{\partial}{\partial x} - \cos\lambda\frac{\partial}{\partial t}\right) f = 0, \\ \left(\frac{\partial}{\partial y} - \sin\lambda\frac{\partial}{\partial t}\right) f = 0, \end{cases}$$ (69)

where CD is f. The wave equation is the enveloping equation of the system (68) (which depends on λ). The analogue of the Radon transform is

$$J(\Box)f(t) = \int f(t\cos\lambda, t\sin\lambda, t)d\lambda.$$ (70)

We can replace $d\lambda$ by any other measure as before. The interesting measures are $\exp(im\lambda)d\lambda$. As mentioned above, we can deal with $J(\Box)$ without the twistors because $\{\lambda\}$ is now compact. In particular $J(\Box)f$ is the solution of the wave equation which is invariant under rotations and whose WD on the t-axis is f for the trivial character and 0 for the other characters.

Remark. These ideas can be extended to varieties V which are fibred by subvarieties V_λ each of which is a covering of a linear variety, the coverings being related to a fixed CP.

The Radon Transform is defined by the geometry of planes in $\mathbb{R}^n$ and the twistor CP is defined by the geometry of subvarieties of V. We present a direct relation between these geometries. This will help clarify the twistor CP since, in some sense, it avoids the enveloping equation qua differential equations. We shall give a formal argument. The rigorization can be accomplished with some difficulty. We want to form the Fourier transform $\widehat{Jf}$ of Jf as given by (2). The Fourier transform $\widehat{Jf}^b$ in b is given by

$$\widehat{Jf}^b = \int e^{-ia\lambda\cdot\hat{b}}\hat{f}(\hat{b})d\lambda.$$ (71)

Now take the Fourier transform in a. There results

$$\widehat{Jf} = \int\int e^{ia\cdot(\hat{a}-\lambda\hat{b})}\hat{f}(\hat{b})da\, d\lambda = \int \delta_{\hat{a}-\lambda\hat{b}}\hat{f}(\hat{b})d\lambda.$$ (72)

This confirms our previous assertion that $\widehat{Jf}$ is the integral over λ of the functions $\hat{f}(\lambda\hat{b}, \hat{b})$ on V_λ.

We can take this idea further. Suppose A, B are groups and Λ is a set for which $a\lambda$ is defined for each $a \in A$ and is an element of B. Assume that Λ acts on both B and A. Then for a "nice" function f on B we can form

$$Jf = \int f(a\lambda b)d\lambda. \tag{73}$$

Here $d\lambda$ is a suitable measure on Λ. The "product" $a\lambda b$ is interpreted as $(a\lambda)b$. We take a Fourier transform on the group B. Call $\rho_{\hat{b}}$ a representation. We have

$$\widehat{Jf}^b = \int \rho_{\hat{b}}(b)f(a\lambda b)d\lambda\alpha b = \int \rho_{\hat{b}}((a\lambda)^{-1})\hat{f}(\hat{b})d\lambda = \int \rho_{\hat{b}}^{-1}(a\lambda)\hat{f}(\hat{b})d\lambda. \tag{74}$$

Now let $\sigma_{\hat{a}}$ be a representation of A. We have to form

$$\int \sigma_{\hat{a}}(a)\rho_{\hat{b}}^{-1}(a\lambda)da. \tag{75}$$

In general this is a complicated integral, even if $A = B$. We do not know how to proceed. However, if A and B are additive groups of matrices (usually Lie groups) say of $m \times \ell$ and $m \times b$ matrices and Λ is a linear space of $\ell \times b$ matrices, then we can make the calculations as follows:

$$\sigma_{\hat{a}}(a) = e^{i\hat{a}\cdot a},$$

$$\rho_{\hat{b}}(b) = e^{i\hat{b}\cdot b}.$$

Here $\hat{a}$ and $\hat{b}$ belong to (the dual spaces of) A and B. Then (75) becomes

$$\int e^{ia\cdot\hat{a}-a\lambda\cdot\hat{b}}da = \int e^{ia\cdot(\hat{a}-\lambda\hat{b})}da = \delta_{\hat{a}-\lambda\hat{b}} \tag{76}$$

as before. We should be careful to note that $\hat{a}$ and $\lambda\hat{b}$ are matrices in $\hat{A}$ so $\delta_{\hat{a}-\lambda\hat{b}}$ is the δ function of $\hat{a} - \lambda\hat{b} = 0$ as a subset of $\hat{A}$ and then it continues to be constant on $\hat{A}^\perp$.

References

[E1] L. Ehrenpreis, *Fourier Analysis in Several Complex Variables*, Wiley–Interscience, New York, 1970.
[E2] L. Ehrenpreis, *The Universality of the Radon Transform*, Oxford University Press, Oxford, UK, 2003.
[E3] L. Ehrenpreis, Three problems at Mount Holyoke, in *Radon Transforms and Tomography*, Contemporary Mathematics 228, American Mathematical Society, Providence, RI, 2001, 123–130.

Analytic and Algebraic Ideas:
How to Profit from Their Complementarity

Alain Yger

LaBAG
Université Bordeaux 1
351 Cours de la Libération
33405 Talence
France
`yger@math.u-bordeaux.fr`, `labag@math.u-bordeaux.fr`

To my friend Carlos on the occasion of his 60*th birthday.*

Summary. Facing algebraic questions from the analytic point of view has been the guide line of the joint work which I have been pursuing for almost twenty years with Carlos Berenstein. Instead of giving an up-to-date state of the art, I will focus on a few key points which still remain to be clarified and indicate a list of prospective developments where the ideas that analysis suggests, combined with multidimensional residue theory, will certainly have to play a major role. I will point out the following crucial fact, namely that in the "dictionary" between the analytic and algebraic points of view in multidimensional residue theory, integral symbols happen to be the analytic substitutes for power series developments in terms of parameters. The Briançon–Skoda theorem, which seems to be a corner stone between constructions inspired by algebraic ideas on one side and by analytic ideas on the other side, will be a leitmotiv in this talk. Such ideas need to be combined in the future with arithmetic aspects we missed up to now and which indeed imply some additional rigidity.

1 Introduction

The concept of Chow ideal, together with the algebraic notion of the integral closure and Briançon–Skoda theorems (that I will present in Section 2), played a crucial role in D. W. Brownawell's approach to the effective nullstellensatz [Bro]. Later on, J. Kollár proposed in [Ko1] an alternative way to attack this problem which was more directly inspired by algebraic geometry. The search for a sharp arithmetic nullstellensatz (which was finally obtained by T. Krick, L. M. Pardo, M. Sombra in [KPS]) was our main motivation with Carlos Berenstein to reinterpret in the setting of multidimensional residue theory integral representation formulas ([BeY4]). Here comes the following crucial fact: in the "dictionary" between the analytic point of view (dealing, for example, with integral representation formulas of the Bochner–Martinelli type as in [BeY2]) and the algebraic point of view (inspired by multidimensional residue theory, as in [BeY4]), integral symbols happen to be the analytic substitutes for power series developments in terms of parameters. As an illustration, I will focus in Section 2 on different approaches of the Briançon–Skoda theorem which seems to be a corner stone in our constructions lying in between analysis and commutative algebra.

The notion of integral closure is also deeply connected with analytic approaches toward intersection theory. For example, given an intersection cycle $C = C_1 \bullet \cdots \bullet C_M$, Lelong numbers related to C or Lojasiewicz exponents for $I(C)$ (see [CyKT]) are connected to the Chow ideal of C and therefore to the integral closure of the ideal generated by the ideals corresponding to the cycles involved in the intersection. Residual currents constructed from Bochner–Martinelli representation formulas appear to play a crucial role in a generalization to improper intersections of the Lelong–Poincaré factorization formula

$$[V(f_1, \ldots, f_m)] = \left(\bigwedge_{j=1}^{m} \bar{\partial} \frac{1}{f_j} \right) \wedge df_1 \wedge \cdots \wedge df_m.$$

The relation with King's formula is now better understood since the recent work of M. Méo [Meo1, Meo2] and M. Andersson [And3] (following [PTY] and [BeY5]). We will point out in Section 2 how such results could be related to improper intersection theory, as introduced in [Tw].

I will conclude with an invitation to pursue such a program, taking more into account some arithmetic aspects it seems we missed up to now. Note, for example, that understanding the Briançon–Skoda theorem from the algebraic point of view is much easier in positive characteristic, or that the theory of exponential polynomials (which initially inspired our joint work with Carlos Berenstein) has been recently revisited starting with p-adic ideas introduced by B. Dwork, N. Katz, and P. Robba and more recently by Y. André ([DGS, Andr]). For example, Ritt's factorization theorem (which plays, together with analytic division formulas and the Ax–Schanuel theorem, a fundamental role in our study of ideals generated by exponential-polynomials [BeY1]) is more rigid in the arithmetic context: in particular, (see [Andr]), the quotient by $z - 1$ of a sum of exponential

$$\sum_j a_j \exp(i\alpha_j z),$$

(with $a_j, \alpha_j \in \overline{\mathbb{Q}}$) which vanishes at $z = 1$ happens to be a sum of exponentials, which of course fails without arithmetic constraints (this leads to an elegant proof of the Lindemann–Weierstrass theorem). There seems to me to be a parallel between the power of reasoning in positive characteristic when understanding in some algebraic way such mysterious division results as the Briançon–Skoda theorem and that of using ideas issued from p-adic analysis to get such rigidity results for exponential polynomials with both algebraic coefficients and frequencies.

2 Chow ideal, integral closure, and the Briançon–Skoda theorem

Let us adopt here a geometric (semilocal) point of view. Let U be a neighborhood of the origin in $\mathbb{C}^n$ and $C = \sum_{j=1}^N \alpha_j Z_j$, $\alpha_j \in \mathbb{N}^*$, be a purely k-dimensional effective cycle with support $|C|$ in U. The *Chow ideal* of the cycle is defined as follows (see [Ko2]): consider all admissible linear projections $\pi : \mathbb{C}^n \longrightarrow \mathbb{C}^{k+1}$ such that the origin is isolated in $\mathrm{Ker}\,\pi \cap |C|$ and the restriction of π to $U \cap |C|$ is proper (in the topological sense); for such a map, let $\mu_{\pi,j}$, $j = 1, \ldots, N$, be the number of sheets of the covering $\pi_{U \cap |Z_j|} : U \cap |Z_j| \longrightarrow \pi(U)$ and $F_\pi := \prod_j (f_{\pi,j} \circ \pi)^{\alpha_j \mu_{\pi,j}}$, where $f_{\pi,j} = 0$ is an irreducible equation for the hypersurface $\pi(|Z_j| \cap U)$ in $\pi(U)$; the Chow ideal $I^{\mathrm{chow}}(C)$ of the effective cycle C is defined as the ideal generated by the F_π, π being such an admissible projection. Such a definition can be extended to nonpurely dimensional cycles ($I^{\mathrm{chow}}(C_1 + C_2) = I^{\mathrm{chow}}(C_1) \cdot I^{\mathrm{chow}}(C_2)$). A key fact relating intersection problems (of geometric nature) and division problems (of algebraic nature) is the following: if $C_1, \ldots, C_M$ are M purely dimensional effective cycles with support in U and $C_1 \bullet \cdots \bullet C_M$ denotes the intersection cycle obtained through the Vogel–Tworzewski construction [Tw], then one has

$$I^{\mathrm{chow}}(C_1 \bullet C_2 \bullet \cdots \bullet C_M) \subset \overline{(I(C_1), \ldots, I(C_M))}, \qquad (1)$$

where the ideal of an effective cycle $C = \sum_j a_j Z_j$ is defined as

$$I(C) := \prod_j (\{f;\ f = 0 \text{ on } |Z_j|\})^{\alpha_j} \qquad (2)$$

and the bar over the right-hand side of this inclusion means the *integral closure* of the ideal. We recall here that the *integral closure* of an ideal I in a noetherian domain R is defined as the ideal $\overline{I}$ of all elements in R which satisfy an homogeneous relation of integral dependency

$$h^N + a_1 h^{N-1} + \cdots + a_N \equiv 0 \qquad (3)$$

with $a_k \in I^k$ for $k = 1, \ldots, N$. Equivalently, $\overline{I}$ consists of all elements h such that $h \in IV$ for all discrete valuation rings V lying between R and its fraction field $(3)'$, or all elements $h \in R$ such that there exists $c \in R^*$ with $ch^n \in I^n$ for infinitely many $n \in \mathbb{N}$ $(3)''$ (see, for example, [Smi] for a survey and an updated list of references about such a notion, together with its relation to the notion of *tight closure* in a noetherian domain with prime characteristic, to which we will come back later).

Summarizing here, we have at least three equivalent ways to assert that some $h \in R$ belongs to the integral closure of I. In the particular case $R = \mathcal{O}_n$, testing the valuative criterion for h amounts to say that, given a system $(f_1, \ldots, f_m)$ of generators for I, one has $|h| \leq K \max |f_j|$ for some constant K in a neighborhood of the origin. Such an analytic characterization will be deeply connected with the use of integral formulas of the Bochner–Martinelli type in order to solve explicitly division problems: generalized versions of the Lelong–Poincaré formulas expressed in terms of such integral formulas lead (when variables are duplicated) to extensions of Lagrange interpolation formulas of the Cauchy–Weil–Bergman type [We]. It remains an interesting challenge, in the case $R = \mathcal{O}_n$, to show (without going through resolution of singularities and the valuative criterion as in [LejT]) that any h satisfying an inequality $|h| \leq K \max |f_j|$ fulfills an algebraic identity of the form (3).

When R is a regular n-dimensional local ring (such as $\mathcal{O}_n$), then a fundamental result (initially proved in $\mathcal{O}_n$ by J. Briançon and H. Skoda in [BriS] using L^2-methods, then extended by J. Lipman and A. Sathaye in the general setting [LiS]) asserts that for any ideal I in R, for any $k \in \mathbb{N}^*$, the integral closure of $I^{k+\mu-1}$ lies in I^k (in particular $\overline{I^\mu} \subset I$ taking $k = 1$) if μ is defined as the minimum of n and of the least number of generators of I. When R has positive characteristic p, the proof of this result (for $k = 1$) follows easily from the rather immediate intermediate inclusions $\overline{I^r} \subset I^* \subset I$, where r denotes the minimal number of generators of I and I^* denotes the tight closure of $I = (f_1, \ldots, f_r)$, that is the set of $h \in R$ such that there exists a nonzero element $c \in R$ with $ch^{p^e} \in (f_1^{p^e}, \ldots, f_r^{p^e})$ for all $e >> 0$. One needs here to use the last equivalent formulation (3)'' we proposed above for the membership to the integral closure [HoH1, Smi]. In the 0-characteristic case, such arguments do not apply directly, but it is fundamental to notice (as pointed out by B. Teissier in [Te3]), that Caratheodory's theorem combines with a generalization of a theorem of Fenchel [HanR] to provide a simple proof of the Briançon–Skoda theorem for monomial ideals. One should add here that the descent from characteristic p down to characteristic zero, which was very recently described in [HoH2], provides indeed some new light on the Briançon–Skoda theorem. It seems important to connect the naïve ideas we discuss here (inspired by Bochner–Martinelli-type integral formulas or by their algebraic counterpart, multidimensional residue theory in its algebraic presentation; see, for example, [Lip]), to such an approach going from characteristic p down to characteristic 0.

Chow ideas (initially developed by Y. Nesterenko) were used in 1987 by D. W. Brownawell who combined them with the Briançon–Skoda theorem in order to state the following striking result.

Theorem ([Bro]). *Let $P_1, \ldots, P_m$ be m polynomials in $\mathbb{C}[X_1, \ldots, X_n]$ such that $\deg P_j \leq D$, $j = 1, \ldots, m$, with no common zeroes in $\mathbb{C}^n$; then one can write a Bézout identity $1 = P_1 Q_1 + \cdots + P_m Q_m$ such that $\max \deg P_j Q_j \leq 3n\mu D^\mu$, where $\mu := \min(n, m)$.*

As a matter of fact, Brownawell's result comes as a combination of Bézout's theorem, the inclusion (1), and finally the Briançon–Skoda theorem which plays the role of a "black box." The following result (which is due to M. Hickel and answers positively a question our analytic point of view with Carlos Berenstein motivated us to ask in [BeY3]) emphasizes the crucial role of the Briançon–Skoda theorem as a path from intersection to division problems (see also [EL] for a related point of view).

Theorem ([Hi]). *Let $P_1, \ldots, P_m$ be m polynomials in $\mathbb{C}[X_1, \ldots, X_n]$ such that $\deg P_j \leq D$, $j = 1, \ldots, m$, and Q lies locally in the integral closure of $(P_1, \ldots, P_n)$ at any point in $\mathbb{C}^n$ (for example, $Q \in (P_1, \ldots, P_m)$). Then one can write an explicit division formula:*

$$Q^{\min(n+1,m)} = \sum_{j=1}^{m} P_j Q_j, \quad \deg P_j Q_j \leq \min(n+1,m)(\deg Q + D^{\min(n,m)}).$$

A key stumbling block with respect to the use of the Briançon–Skoda theorem (or its natural analytic companion, the Bochner–Martinelli integral representation formula) when one wants to keep track of algebraic or arithmetic constraints while writing explicit division formulas is the fact that L^2-methods (or complex analytic methods involving integral representation formulas), usually hide the algebraic argument.

In 1981, J. Lipman and B. Teissier proposed in [LiT] an alternative proof of the Briançon–Skoda theorem based on the use of multidimensional theory (through integral symbols): let $f_1, \ldots, f_n$ define an $\mathfrak{m}$-primary ideal in $\mathcal{O}_n$ ($\mathfrak{m}$ being the maximal ideal in the local ring $\mathcal{O}_n$) and h lying in the integral closure of $(f_1, \ldots, f_n)$; then for any $g \in \mathcal{O}_n$, the Cauchy-like integral

$$\frac{1}{(2i\pi)^n} \int_{\Gamma_f(\epsilon_1, \ldots, \epsilon_n)} \frac{h^n g \, d\zeta_1 \wedge \cdots \wedge d\zeta_n}{f_1 \cdots f_n}, \tag{4}$$

where $\Gamma_f(\epsilon) := \{|f_1| = \epsilon_1, \ldots, |f_n| = \epsilon_n\}$ (which is constant for ϵ generic thanks to Stokes's theorem) tends to 0 when ϵ tends to zero (remaining in a cone). Therefore (4) is equal to zero, which implies $h^n \in (f_1, \ldots, f_n)$ because of the duality theorem for Grothendieck residue symbols attached to regular sequences. The algebraic counterpart of the Cauchy integral symbol is the Cauchy development. It is important to recall here that, for such $f_1, \ldots, f_n$ and $r \in \mathcal{O}_n$, the residue symbol

$$\mathrm{Res} \begin{bmatrix} r \, d\zeta_1 \wedge \cdots \wedge d\zeta_n \\ f_1, \ldots, f_n \end{bmatrix} = \lim_{\epsilon \to 0} \frac{1}{(2i\pi)^n} \int_{\Gamma_f(\epsilon_1, \ldots, \epsilon_n)} \frac{r \, d\zeta_1 \wedge \cdots \wedge d\zeta_n}{f_1 \cdots f_n}$$

is not defined alone in algebraic presentations of multidimensional residue theory (as, for example, in [Lip, Chapter 3]). Such a residue symbol comes as the trace of an operator $T \in \mathrm{Hom}_{\mathbb{C}}(\mathcal{O}_n/(f_1, \ldots, f_n), \mathcal{O}_n/(f_1, \ldots, f_n))$. It is defined together with the whole list of symbols

$$\mathrm{Res} \begin{bmatrix} r \, d\zeta_1 \wedge \cdots \wedge d\zeta_n \\ f_1^{k_1+1}, \ldots, f_n^{k_n+1} \end{bmatrix}, \quad (k_1, \ldots, k_n) \in \mathbb{N}^n.$$

Note that, in the same vein, the Briançon–Skoda theorem for $I = (f_1, \ldots, f_n)$ is not only the inclusion $\overline{I^n} \subset I$, but the whole list of inclusions $\overline{I^{n+k-1}} \subset I^k$ for any $k \in \mathbb{N}^*$. Here is an alternative way to relate such a list of inclusions (in terms of Taylor developments instead of Cauchy integrals) to the duality theorem.

A Rephrasing of the Briançon–Skoda Theorem. *Let $(f_1, \ldots, f_n)$ and $(g_1, \ldots, g_n)$ be two $\mathfrak{m}$-primary ideals in $\mathcal{O}_n$ with the same integral closure; then the formal power series (in $\mathbb{C}[[u]]$)*

$$\sum_{k_1=0}^{\infty} \cdots \sum_{k_n=0}^{\infty} \mathrm{Res} \begin{bmatrix} r g_1^{k_1+1} \cdots g_n^{k_n+1} d\zeta_1 \wedge \cdots \wedge d\zeta_n \\ f_1^{k_1+1}, \ldots, f_n^{k_n+1} \end{bmatrix} u_1^{k_1} \cdots u_n^{k_n}$$

is identically zero for any $r \in \mathcal{O}_n$.

In order to justify such a reformulation, one should recall here the key remark of M. Hochster in the appendix to [LiT]: given a regular sequence $(f_1, \ldots, f_n)$ in $\mathcal{O}_n$, one has, for any $k \in \mathbb{N}^*$,

$$(f_1, \ldots, f_n)^k = \bigcap_{\substack{l \in \mathbb{N}^n \\ l_1 + \cdots + l_n = k-1}} (f_1^{l_1+1}, \ldots, f_n^{l_n+1}). \tag{5}$$

Besides the three ways we already mentioned to characterize the membership of a given $h \in \mathcal{O}_n$ to the integral closure of some m-primary ideal $(f_1, \ldots, f_n)$, let us formulate here two alternative ones involving multidimensional residue calculus. Both of them appear to be deeply connected with the various techniques we have been developing with Carlos Berenstein (culminating in [BeY4]) in order to make "effective" the black box which corresponds to the involvement of the Briançon–Skoda theorem in our work.

Characterization 1 ([TsiY, Theorem 4.10]). *An element $h \in \mathcal{O}_n$ belongs to the integral closure of $(f_1, \ldots, f_n)$ if and only if there exists $N \in \mathbb{N}$ such that for any $r \in \mathcal{O}_n$, the power series*

$$\sum_{k_1=0}^{\infty} \cdots \sum_{k_n=0}^{\infty} \mathrm{Res} \left[\frac{r h^{k_1 + \cdots + k_n} d\zeta_1 \wedge \cdots \wedge d\zeta_n}{f_1^{k_1+1}, \ldots, f_n^{k_n+1}} \right] u_1^{k_1} \cdots u_n^{k_n}$$

corresponds to the development about $u = 0$ of a rational function F_1/F_2 with no pole at $u = 0$, F_2 independent of r and $\max(\deg F_1, \deg F_2) \leq N$.

Characterization 2. *An element $h \in \mathcal{O}_n$ belongs to the integral closure of $(f_1, \ldots, f_n)$ if and only if there exists $N \in \mathbb{N}$ and $c \in \mathcal{O}_n$ such that for any $r \in \mathcal{O}_n$, the formal power series*

$$\sum_{k_1=0}^{\infty} \cdots \sum_{k_n=0}^{\infty} \mathrm{Res} \left[\frac{r c h^{k_1 + \cdots + k_n + 1} d\zeta_1 \wedge \cdots \wedge d\zeta_n}{f_1^{k_1+1}, \ldots, f_n^{k_n+1}} \right] u_1^{k_1} \cdots u_n^{k_n}$$

is a polynomial in u with degree less than N.

Such a characterization is obtained from characterization $(3)''$ using (5) and the duality theorem.

In order to unify the picture (and understand completely the reason why multivariate residue calculus appears as an algebraic counterpart of integral representation formulas or L^2 methods), it would be crucial to derive a direct "combinatoric" proof of the Briançon–Skoda theorem which does not use (even in some artificial way) an integral symbol such as the integration on the chain $\Gamma_f(\epsilon)$ in (4). Unfortunately, despite many efforts with M. Hickel, we could not achieve such an objective. Let us sketch briefly what could be a conjectural scheme of the proof: first start with two m-primary ideals $(f_1, \ldots, f_n)$ and $(g_1, \ldots, g_n)$ with the same integral closure. Then introducing additional parameters $u_1, \ldots, u_n$ and noting that formally (this follows from the transformation law in residue calculus; see, for example, [GrH, Chapter 6]), one has, for any $r \in \mathcal{O}_n$,

$$
\sum_{k_1=0}^{\infty} \cdots \sum_{k_n=0}^{\infty} \operatorname{Res} \left[\begin{matrix} r g_1^{k_1+1} \cdots g_n^{k_n+1} d\zeta_1 \wedge \cdots \wedge d\zeta_n \\ f_1^{k_1+1}, \ldots, f_n^{k_n+1} \end{matrix} \right] u_1^{k_1} \cdots u_n^{k_n}
$$

$$
= \operatorname{Res} \left[\begin{matrix} r g_1 \cdots g_n d\zeta_1 \wedge \cdots \wedge d\zeta_n \\ f_1 - g_1 u_1, \ldots, f_n - g_n u_n \end{matrix} \right]
$$

$$
= \frac{(-1)^n}{(u_1 \cdots u_n)^2} \operatorname{Res} \left[\begin{matrix} r g_1 \cdots g_n d\zeta_1 \wedge \cdots \wedge d\zeta_n \\ g_1 - f_1/u_1, \ldots, g_n - f_n/u_n \end{matrix} \right].
$$

Reversing the roles of f and g and introducing additional parameters $v_1, \ldots, v_n$ deduce that

$$
\sum_{k_1=0}^{\infty} \cdots \sum_{k_n=0}^{\infty} \operatorname{Res} \left[\begin{matrix} r f_1^{k_1+1} \cdots f_n^{k_n+1} d\zeta_1 \wedge \cdots \wedge d\zeta_n \\ g_1^{k_1+1}, \ldots, g_n^{k_n+1} \end{matrix} \right] v_1^{k_1} \cdots v_n^{k_n}
$$

$$
= \operatorname{Res} \left[\begin{matrix} r g_1 \cdots g_n d\zeta_1 \wedge \cdots \wedge d\zeta_n \\ g_1 - f_1 v_1, \ldots, g_n - f_n v_n \end{matrix} \right].
$$

On the other hand, thanks to Characterization 2, the fact that $g_1, \ldots, g_n$ lie in $(f_1, \ldots, f_n)$ implies that there exists a nonzero element $c \in \mathcal{O}_n$ and $N \in \mathbb{N}$ such that, for any $r \in \mathcal{O}_n$, the formal power series

$$
\sum_{k_1=0}^{\infty} \cdots \sum_{k_n=0}^{\infty} \operatorname{Res} \left[\begin{matrix} r c g_1^{k_1+1} \cdots g_n^{k_n+1} d\zeta_1 \wedge \cdots \wedge d\zeta_n \\ f_1^{k_1+1}, \ldots, f_n^{k_n+1} \end{matrix} \right] u_1^{k_1} \cdots u_n^{k_n}
$$

is a polynomial in u with degree N. Since the polar set of the rational function whose development about the origin is

$$
\sum_{k_1=0}^{\infty} \cdots \sum_{k_n=0}^{\infty} \operatorname{Res} \left[\begin{matrix} r g_1^{k_1+1} \cdots g_n^{k_n+1} d\zeta_1 \wedge \cdots \wedge d\zeta_n \\ f_1^{k_1+1}, \ldots, f_n^{k_n+1} \end{matrix} \right] u_1^{k_1} \cdots u_n^{k_n}
$$

does not depend on r, there is some hope to show that one get get rid of c and get that such a rational function is in fact a polynomial in u, which contradicts (if it is not identically zero) the fact that it could be also expressed as a polynomial in $1/u_1, \ldots, 1/u_n$ (as expected through the transformation law).

   Let us go now from the local setting to the global setting and replace the local ring $\mathcal{O}_n$ with the polynomial ring $\mathbb{C}[X_1, \ldots, X_n]$. Our long term work with Carlos Berenstein convinced me that what makes the algebraic and analytic approaches so different (and therefore so complementary) is that there are two different perceptions of the hyperplane at infinity when dealing with one point of view or the other. In algebraic geometry, the hyperplane at infinity is nothing else than a projective hyperplane (it can be any projective hyperplane after a projective change of coordinates).

For the analyst, who looks at algebraic division problems from the affine space $\mathbb{C}^n$, the notion of infinity makes sense and carries related notions such as, for example, the notion of topological properness for a polynomial map from $\mathbb{C}^n$ to $\mathbb{C}^m$, the concept of *separation at infinity*, together with the related definition of the *Lojasiewicz exponent at infinity* (see, for example, [CyKT] for a general overview and references about these fundamental notions which have been extensively studied in the past years by the Crakow school), the concept of *apparent contour* which was introduced by G. Monge, etc.

Then let $P = (P_1, \ldots, P_m)$ be a polynomial map from $\mathbb{C}^n$ to $\mathbb{C}^m$ such that the zero set $V(P_1, \ldots, P_n) := \{\zeta \in \mathbb{C}^n; \ P(\zeta) = 0\}$ is discrete (hence finite). Let $\mathfrak{I}$ be the ideal generated in $\mathbb{C}[X_0, \ldots, X_n]$ by the homogenizations of the P_j; let $V(\mathfrak{I})$ be the zero set of $\mathfrak{I}$ in $\mathbb{P}^n(\mathbb{C})$. One can define (see [Hi]) in terms of the polar invariants attached (see [Te1]) to the normalized blow-up of $\mathbb{P}^n(\mathbb{C})$ with center $V(\mathfrak{I})$, the least positive rational number $\nu_\infty(\mathfrak{I})$ such that

$$\sum_{j=1}^{m} \frac{P_j(\zeta)}{\|\zeta\|^{\deg P_j}} \geq \kappa \|\zeta\|^{-\nu_\infty(\mathfrak{I})}$$

for some $\kappa > 0$ when $\|\zeta\| >> 0$. When $\nu_\infty(\mathfrak{I}) < \deg P_j$ for $j = 1, \ldots, m$, the polynomial map P is proper in the topological sense; the case $\nu_\infty(\mathfrak{I}) = 0$ corresponds to the particular case when the supports of the projective divisors corresponding to $P_1, \ldots, P_m$ do not intersect on the hyperplane at infinity. The following result (which was transposed in [VY] from the projective setting to the toric setting [GeKZ]) appears [BoH] to be the corner stone of the effective results [BeY2, BeY4] that gave the first result toward the formulation of an effective arithmetic nullstellensatz.

Theorem ([VY]). *Assume $(P_1, \ldots, P_n)$ is a proper map from $\mathbb{C}^n$ to $\mathbb{C}^n$. Then for any $Q \in \mathbb{C}[X_1, \ldots, X_n]$ such that*

$$\deg Q < \sum_{j=1}^{n} \deg P_j - n(1 + \nu_\infty(\mathfrak{I})) = \sum_{j=1}^{n} (\deg P_j - \nu_\infty(\mathfrak{I})) - n,$$

the total residue symbol (that is the sum of local Grothendieck residues at all points in $V(P_1, \ldots, P_n)$)

$$\mathrm{Res} \left[\begin{array}{c} Q(X) dX_1 \wedge \cdots \wedge dX_n \\ P_1, \ldots, P_n \end{array} \right]$$

is equal to 0.

This result can be proved using the Bochner–Martinelli integral representation formula [VY]. M. Hickel noticed that an alternative way to prove it is to use the Briançon–Skoda theorem in its full strength, that is, for any ideal $\mathfrak{I}$ generated by n elements in $\mathcal{O}_{n+1}$ and for any $k \in \mathbb{N}^*$, $\overline{\mathfrak{I}^{k+n-1}} \subset \mathfrak{I}^k$ for any $k \in \mathbb{N}^*$ (note that the assertion for $k = 1$ is not sufficient, which shows again that integral symbols involved in Bochner–Martinelli integral formulas hide some Cauchy-like development). One should add here that a currential geometric reinterpretation of this result was recently proposed in [And1].

Because of its ubiquity, the Briançon–Skoda theorem provides a dictionary between analytic information (such as Lojasiewicz *inequalities*) and algebraic information (for example, the vanishing of a formal power series defined in terms of residue symbols). Here is an example of such a result.

Proposition ([TsiY, Lemma 4.2]). *Let* $P_1, \ldots, P_n \in \mathbb{C}[X_1, \ldots, X_n]$ *be a proper polynomial map from* $\mathbb{C}^n$ *to* $\mathbb{C}^n$*; the Lojasiewicz exponent at infinity of P is at most* 1 *(that is* $\liminf_{\|\varsigma\| \to \infty} \|P(\varsigma)\|/\|\varsigma\|^{1-\epsilon} > 0$ *for any* $\epsilon > 0$*, which is clearly an analytic assertion) if and only if all residue symbols*

$$\mathrm{Res} \begin{bmatrix} r X_1^{l_1} \cdots X_n^{l_n} dX_1 \wedge \cdots \wedge dX_n \\ P_1^{k_1+1}, \ldots, P_n^{k_n+1} \end{bmatrix}, \quad l_1 + \cdots + l_n < k_1 + \cdots + k_n$$

are equal to zero (this is a list of algebraic relations).

3 Multiplying integration currents

At the same time we pushed forward our efforts to clarify the dictionary between analytic tools (such as integral representation formulas) and algebraic ones (multivariate residue calculus adding parameters as described in Section 2), we went on profiting from analytic aspects of multidimensional residue theory, dealing with the much less rigid *currential* point of view [Lel]. I will just point out here relevant recent results with potential applications since they are closely related either to the integral closure and Briançon–Skoda theorems or to intersection theory. We were interested in both questions with Carlos Berenstein when profiting from the complementarity of analytic, geometric and algebraic ideas. For further references and a complete up-to-date overview of all such topics, I refer to the two survey papers [Bjo] and [TsiY].

Let U be some open set in $\mathbb{C}^n$, $F_1, \ldots, F_M$, M holomorphic functions in U. Exploring analytic ideas involved in multivariate residue calculus and inspired (as the analytic ideas lying behind the concept of algebraic closure introduced in the previous section) by normalized blowing-up techniques and toric geometry, we introduced with M. Passare, A. Tsikh, and Carlos Berenstein in [Y1, Y2, PTY, BeY4] a list of currents $T_{\mathcal{I}}(F, \mathfrak{p})$ which is indexed by the collection of all nonempty ordered subsets $\mathcal{I} = \{i_1, \ldots, i_k\}$ (with $1 \le i_1 < \cdots < i_k \le M$) of $\{1, \ldots, M\}$ and by the collection of multiindexes $\mathfrak{p} = (\mathfrak{p}_1, \ldots, \mathfrak{p}_M)$ in $\mathbb{N}^M$. For $k \in \{1, \ldots, \min(M, n)\}$ and $\mathcal{I}$ with cardinal k, for each $\mathfrak{p} \in \mathbb{N}^M$, the "residual current" $T_{\mathcal{I}}(F, \mathfrak{p})$ is the $(0, k)$ current whose action on an $(n, n-k)$-test form φ can be expressed as

$$\langle T_{\mathcal{I}}(F, \mathfrak{p}), \varphi \rangle := \lim_{\epsilon \to 0} \frac{(-1)^{k(k-1)/2}(k-1)!}{(2i\pi)^k} \frac{1}{\epsilon^k} \int_{\|F\|_{\mathfrak{p}}^2 = \epsilon} \Omega(F; \mathcal{I}; \mathfrak{p}) \wedge \varphi,$$

where

$$\|F\|_{\mathfrak{p}}^2 := \sum_{j=1}^{M} |F_j|^{2(\mathfrak{p}_j + 1)},$$

$$\Omega(F; \mathcal{I}; \mathfrak{p}) := \sum_{l=1}^{k} (-1)^{l-1} \overline{F_{i_l}} \bigwedge_{\substack{\nu=1 \\ \nu \neq l}}^{k} d[\overline{F_{i_\nu}} |F_{i_\nu}|^{2\mathfrak{p}_{i_\nu}}].$$

The current $T_{\mathcal{I}}(F, \mathfrak{p})$ with $\#\mathcal{I} = k$, $k = 1, \ldots, \min(n, M)$, is annihilated in U (as a current) by holomorphic functions h which locally satisfy about any point in U the condition

$$\left(\prod_{i \in \mathcal{I}} F_i^{\mathfrak{p}_i} \right) h \in \overline{(F_1^{\mathfrak{p}_1+1}, \ldots, F_M^{\mathfrak{p}_M+1})^{\#\mathcal{I}}}.$$

Moreover, the support of all such currents (for any subset $\mathcal{I}$, for any multiindex $\mathfrak{p} \in \mathbb{N}^n$) lies in the closed analytic set $V(F) = \{z \in U; F(z) = 0\}$ since it is locally annihilated (as a current) by any antiholomorphic function which vanishes on $V(F)$. From such a property, one could guess, at least heuristically, that the action of the residual current $T_{\mathcal{I}}(F, \mathfrak{p})$ involves only the holomorphic differential operators $\partial/\partial z_j$ for $j = 1, \ldots, n$. Though they fail to be $\bar{\partial}$-closed (except in the particular case $k = M \leq n$), such currents $T_{\mathcal{I}}(F, \mathfrak{p})$ would probably gain to be better understood from the algebraic point of view.

Let C be an effective analytic cycle with support in U and defining ideal (multiplicities being taken into account as in (2)) $I(C) = (f_1, \ldots, f_m)$. Let also $S := \{L_1 = \cdots = L_r = 0\}$ be an s-dimensional reduced submanifold in U. In [Tw], P. Tworzewski introduced (through an algorithm inspired by the Vogel–Stückrad construction) the local *multiindex* of contact $\nu(C, S) = (\nu_s(C, S), \ldots, \nu_0(C, S)) \in \mathbb{N}^{s+1}$ between the cycle C and the smooth s-dimensional submanifold S at some point x lying in the intersection of the support of C and S. He also introduced the related intersection cycle $C \bullet S$. It was shown by Achilles and Rams [AchR] (see also [AchM]) that the components of the multiindex of contact $\nu \in \mathbb{N}^{n-r+1}$ can be understood as generalized Hilbert–Samuel multiplicities.

On the other hand, let $M = r + m$ and $F := (L_1, \ldots, L_r, f_1, \ldots, f_m)$. For any $k \in \{0, s\}$, for any $p \in \mathbb{N}^r$, the currents $T_{\mathcal{I}}(F; \mathfrak{p})$, with $\#\mathcal{I} = n - k$ (where $\mathfrak{p} := (p_1, \ldots, p_r, 0, \ldots, 0)$) can be combined with the $dF_1, \ldots, dF_M$ in order to realize the positive ∂ and $\bar{\partial}$-closed $(n - k, n - k)$ current

$$[V_{\mathfrak{p}}(F)]_k := \sum_{\substack{\mathcal{I} \\ \#\mathcal{I}=n-k}} \prod_{i \in \mathcal{I}} (\mathfrak{p}_i + 1) T_{\mathcal{I}}(F, \mathfrak{p}) \wedge dF_{i_1} \wedge \cdots \wedge dF_{i_{n-k}}$$

(see [BeY5]). When $k = s$, it follows from King's formula [King] that at least if the p_j are large enough, the current $[V_{\mathfrak{p}}(F)]_s$ vanishes if $\dim(S \cap V(f_1, \ldots, f_m)) < s$. Otherwise (see [BeY5, Meo1, Meo2, And3]), S is involved as an irreducible component in the decomposition of the analytic subset $S \cap V(f_1, \ldots, f_m)$ and the current $[V_{\mathfrak{p}}(F)]_s$ equals Z_s, where Z_s is the s-dimensional component of the intersection cycle $C \bullet S$ between C and the smooth manifold S (the local multiplicity at z of such a cycle Z_s being the first component in the multiindex of intersection between C and S at the point z in the sense of [Tw]). When $k = 0, \ldots, s - 1$, the closed positive current $[V_{\mathfrak{p}}(F)]_k$ admits a unique Siu decomposition

$$[V_{\mathfrak{p}}(F)]_k = \sum_j \alpha_{\mathfrak{p},k,j} C_{\mathfrak{p},k,j} + R_{\mathfrak{p},k}$$

(see [Siu, Dem1, Dem2]) and it seems a natural question to ask whether, for $p_1, \ldots, p_r$ large enough (or if not for which p) the singular part

$$\sum_j \alpha_{\mathfrak{p},k,j} C_{\mathfrak{p},k,j}$$

fits with the k-dimensional component in the decomposition of the intersection cycle $C \bullet S$ (as defined by Tworzewski in [Tw]). This would give some way to approximate the Tworzewski multiindex of contact in terms of the Bochner–Martinelli integral representation formulas (as happens, of course, when C and S intersect properly).

Such factorization results of the Lelong–Poincaré type for integration currents in the improper intersection case enhance the role of multidimensional residue theory when multiplying in a robust way (as described in [Bjo]) integration currents with meromorphic forms (either in a product space as for the construction of Green currents [BeY5] or on an incidence manifold as far as the concept of trace is concerned). I will just refer here to our recent work [BeVY] and to the "algebraic" formulations (in terms of the rigidity of a nonlinear, nonhomogeneous system of differential equations whose solutions are all automatically rational when the right-hand side is) of Abel's inverse theorem (see [Gr, HeP, Y3, Wei]). In the same vein I would like also to mention before concluding this section the recent work of B. Fabre [Fab], extending Abel's theorem replacing the notion of trace of a meromorphic form by that of Abel–Radon transform of a residual current.

4 Some conclusive and prospective remarks

We have seen how much simpler the understanding of results such as the Briançon–Skoda theorem could be when working in positive characteristic (where the concept of tight closure happens to be a quite useful intermediate notion toward the notion of integral closure). What we tried to point out in this short, quite oriented survey is that any time one wants to substitute an algebraic argument to the use of an integral representation, one has to add additional parameters and study the algebraic properties of formal power series involving residual symbols: do such power series correspond to polynomials, developments of rational functions, of algebraic functions, etc.? The sequence of coefficients of such power series is algebraically governed by transformation laws involved in residue calculus (or generalized versions of transformation laws, as in [BeY4]).

Rigidity constraints that lead to criteria of rationality in $K[[u]]$, where K is a field, imply usually much stronger results in positive characteristic or under arithmetic conditions that allow use of p-adic analysis (working over a number field). For example, the combination of Katz and Chudnowsky theorems (as in [DGS, Andr]) is fundamental to conclude the rationality of solutions of some peculiar differential systems $D - \Gamma$ in $K[[X]]^\mu$, where $\Gamma \in \mathcal{M}_{\mu,\mu}(K[X])$, K being a number field.

There seems to be some interest in transposing such ideas (inspired by p-adic analysis) to the range of methods we summarized in this very partial presentation. Note that the search for Bernstein–Sato-type identities, which is central in our attack with Carlos Berenstein on division problems (division being replaced by integration by parts), could also benefit from such a context. This is probably what we missed in our previous study of ideals generated by exponential polynomials. I cannot conclude this survey without mentioning that L. Ehrenpreis's question (are the zeroes of a sum of exponentials with algebraic coefficients and frequencies well separated?) was (and still remains) a central motivation for introducing all such techniques; a positive answer to such a question is conditioned by the existence of Bernstein–Sato identities (thinking about analytic division formulas as a tool to attack the problem) or by a better understanding of multidimensional residue calculus (its algebraic companion) in positive characteristic or over a number field (where p-adic ideas can be used).

References

[AchM]   R. Achilles and M. Manaresi, Multiplicity for ideals of maximal analytic spread and intersection theory, *J. Math. Kyoto Univ.*, **33**-4 (1993), 1029–1046.

[AchR]   R. Achilles and S. Rams, Intersection numbers, Segre numbers and generalized Samuel multiplicities, *Arch. Math.*, **77** (2001), 391–398.

[And1]   M. Andersson, Residue currents and ideals of holomorphic functions, *Bull. Sci. Math.*, **128**-6 (2004), 481–512.

[And2]   M. Andersson, *The Membership Problem for Polynomial Ideals in Terms of Residue Currents*, preprint, 2004.

[And3]   M. Andersson, Residues of holomorphic sections and Lelong currents, in *Ark. Mat.*, to appear.

[Andr]   Y. André, Séries Gevrey de type arithmétique II: Transcendance sans transcendance, *Ann. Math.*, **151** (2000), 741–756.

[BeGVY]   C. A. Berenstein, R. Gay, A. Vidras, and A. Yger, *Residue Currents and Bezout Identities*, Progress in Mathematics 114, Birkhäuser, Boston, 1993.

[BeY1]   C. A. Berenstein and A. Yger, Ideals generated by exponential-polynomials, *Adv. Math.*, **60** (1986), 69–120.

[BeY2]   C. A. Berenstein and A. Yger, Effective Bézout identities in $\mathbb{Q}[z_1, \ldots, z_n]$, *Acta Math.*, **166** (1991), 69–120.

[BeY3]   C. A. Berenstein and A. Yger, Formules de représentation intégrale et problèmes de division, in P. Philippon, ed., *Diophantine Approximations and Transcendental Numbers, Luminy* 1990, de Gruyter, New York, 1992, 15–37.

[BeY4]   C. A. Berenstein and A. Yger, Residue calculus and effective Nullstellensatz, *Amer. J. Math.*, **121**-4 (1999), 723–796.

[BeY5]   C. A. Berenstein and A. Yger, Residue currents, integration currents in the non-complete intersection case, *J. Reine. Angew. Math.*, **527** (2000), 203–235.

[BeVY]   C. A. Berenstein, A. Vidras, and A. Yger, Analytic residues along algebraic cycles, *J. Complexity*, **21**-1 (2005), 5–42.

[Bjo]   J. E. Björk, Residues and $\mathcal{D}$-modules, in O. A. Laudal and R. Piene, *The Legacy of Niels Henrik Abel: The Abel Bicentennial, Oslo* 2002, Springer-Verlag, New York, 2004, 605–652.

[BoH]   J. Y. Boyer and M. Hickel, Extension dans un cadre algébrique d'une formule de Weil, *Manuscripta Math.*, **98** (1999), 1–29.

[BriS]   J. Briançon and H. Skoda, Sur la clôture intégrale d'un idéal de germes de fonctions holomorphes en un point de $\mathbb{C}^n$, *C. R. Acad. Sci. Paris Sér.* A, **278** (1974), 949–951.

[Bro]   D. W. Brownawell, Bounds for the degrees in the Nullstellensatz, *Ann. Math.*, **126** (1987), 577–592.

[CyKT]   E. Cygan, T. Krasiński, and P. Tworzewski, Separation of algebraic sets and the Lojasiewicz exponent of a polynomial mapping, *Invent. Math.*, **136**-1 (1999), 75–87.

[Dem1]   J.-P. Demailly, Courants positifs et théorie de l'intersection, *Gaz. Math.*, **53** (1992), 131–159.

[Dem2]   J.-P. Demailly, Monge–Ampère operators, Lelong numbers and intersection theory, in *Complex Analysis and Geometry*, University Series in Mathematics, Plenum, New York, 1993, 115–193.

[DGS]   B. Dwork, G. Gerotto, and F. J. Sullivan, *An introduction to G-Functions*, Annals of Mathematics Studies 133, Princeton University Press, Princeton, NJ, 1994.

[EL]   L. Ein and R. Lazarsfeld, A geometric effective nullstellensatz, *Invent. Math.*, **137** (1999), 427–448.

[Fab]   B. Fabre, Sur la transformation d'Abel–Radon des courants localement résiduels, *C. R. Math. Acad. Sci. Paris*, **338**-10 (2004), 787–792.

[GeKZ]   I. M. Gelfand, M. M. Kapranov, and A. V. Zelevinsky, *Discriminants, Resultants and Multidimensional Determinants*, Birkhäuser, Boston, 1994.

[Gr]   P. Griffiths, Variations on a theorem of Abel, *Invent. Math.*, **35** (1976), 321–390.

[GrH]   P. Griffiths and J. Harris, *Principles of Algebraic Geometry*, Wiley–Interscience, New York, 1978.

[HanR]   O. Hanner and H. Radström, A generalization of a theorem of Fenchel, *Proc. Amer. Math. Soc.*, **2** (1951), 589–598.

[HeP]   G. Henkin and M. Passare, Abelian differentials on singular varieties and variation on a theorem of Lie–Griffiths, *Invent. Math.*, **135** (1999), 297–328.

[Hi]   M. Hickel, Solution d'une conjecture de C. Berenstein-A. Yger et invariants de contact à l'infini, *Ann. Inst. Fourier Grenoble*, **51**-3 (2001), 707–744.

[HoH1]   M. Hochster and C. Huneke, Tight closure, invariant theory, and the Briançon–Skoda theorem, *J. Amer. Math. Soc.*, **3** (1990), 31–116.

[HoH2]   M. Hochster and C. Huneke, *Tight Closure in Equal Characteristic Zero*, preprint.

[King]   J. King, A residue formula for complex subvarieties, in *Proceedings of the Carolina Conference on Holomorphic Mappings and Minimal Surfaces*, University of North Carolina at Chapel Hill, Chapel Hill, NC, 1970, 43–56.

[Ko1]   J. Kollár, Sharp effective nullstellensatz, *J. Amer. Math. Soc.*, **1** (1988), 963–975.

[Ko2]   J. Kollár, Effective Nullstellensatz for arbitrary ideals, *J. European Math. Soc.*, **1** (1999), 313–337.

[KPS]   T. Krick, L. M. Pardo, and M. Sombra, Sharp estimates for the arithmetic Nullstellensatz, *Duke Math. J.*, **109**-3 (2001), 521–598.

[LejT]   M. Lejeune-Jalabert and B. Teissier, *Clôture intégrale des idéaux et équisingularité*, Séminaire Lejeune-Teissier, Publications Institut Fourier, Grenoble, 1974.

[Lel]   P. Lelong, *Fonctions plurisousharmoniques et formes différentielles positives*, Gordon and Breach, New York, 1968.

[Lip]   J. Lipman, *Residues and Traces of Differential Forms via Hochschild Homology*, Contemporary Mathematics 61, American Mathematical Society, Providence, RI, 1987.

[LiS]   J. Lipman and A. Sathaye, Jacobian ideals and a theorem of Briançon–Skoda, *Michigan Math. J.*, **28** (1981), 199–222.

[LiT]   J. Lipman and B. Teissier, Pseudo-rational local rings and a theorem of Briançon–Skoda about integral closures of ideals, *Michigan Math. J.*, **28** (1981), 97–116.

[Meo1]   M. Méo, Résidus dans le cas non nécessairement intersection complète, *C. R. Acad. Sci. Paris Sér.* I *Math.*, **333**-1 (2001), 33–38.

[Meo2]   M. Méo, *Courants résidus et formule de King*, preprint.

[PTY]   M. Passare, A. Tsikh, and A. Yger, Residue currents of the Bochner–Martinelli type, *Publ. Mat.*, **44** (2000), 85–117.

[Siu]   Y. T. Siu, Analyticity of sets associated to Lelong numbers and the extension of closed positive currents, *Invent. Math.*, **27** (1974), 53–156.

[Smi]   K. E. Smith, An introduction to tight closure, in *Geometric and Combinatorial Aspects of Commutative Algebra, Messina* 1999, Lecture Notes in Pure and Applied Mathematics 217, Marcel Dekker, New York, 2001, 353–377.

[Te1]   B. Teissier, Variétés polaires II, in *Algebraic Geometry, La Rabida*, Lecture Notes in Mathematics 961, Springer-Verlag, 1980, 71–146.

[Te2]   B. Teissier, Résultats récents d'algèbre commutative effective, in *Séminaire Bourbaki* 1989–1990, Astérisque 189–190, Société Mathématique de France, Paris, 107–131.

[Te3]   B. Teissier, Monomial ideals, binomial ideals, polynomial ideals, in *Proceedings of the Introductory Workshop in Commutative Algebra, September* 2002, Mathematical Sciences Research Institute, Berkeley, CA, to appear.

[TsiY]   A. Tsikh and A. Yger, Residue currents, *J. Math. Sci.*, **120**-6 (2004), 1916–1971.

[Tw]   P. Tworzewski: Intersection theory in complex analytic geometry, *Ann. Polon. Math.*, **62** (1995), 177–191.

[VY]   A. Vidras and A. Yger, On some generalizations of Jacobi's residue formula, *Ann. Sci. École Norm. Sup.*, **34** (2001), 131–157.

[We]   A. Weil, L'intégrale de Cauchy et les fonctions de plusieurs variables, *Math. Ann.*, **111** (1935), 178–182.

[Wei]   M. Weimann, La trace via le calcul résiduel: Une nouvelle version du théorème d'Abel inverse; formes abéliennes, preprint math.CV/0405591, 2004.

[Y1]   A. Yger, Résidus, Courants résiduels et courants de Green, in F. Norguet, S. Ofman, and J.-J. Szczeciniarz, eds., *Géométrie Complexe*, Collection Actualités Scientifiques et Industrielles 1438, Hermann, Paris, 1996, 123–147.

[Y2]   A. Yger, Aspects opérationnels de la théorie des résidus hors du cadre intersection complète, in *Géométrie Complexe* II, Hermann, Paris, 2004, 182–204.

[Y3]   A. Yger, *Abel's Theorem Revisited through New Developments in Residue Theory and Effectivity*, preprint.

On Certain First-Order Partial Differential Equations in $\mathbf{C}^n$

Carlos A. Berenstein[1] and Bao Qin Li[2]

[1] Department of Mathematics
University of Maryland
College Park, MD 20742
USA
carlos@math.umd.edu
[2] Department of Mathematics
Florida International University
Miami, FL 33199
USA
libaoqin@fiu.edu

Summary. The paper is concerned with description of entire solutions in $\mathbf{C}^n$ to first-order partial differential equations of the form $\frac{\partial w}{\partial z_j} = p(z) f(w)$, where p and f are entire or meromorphic functions in $\mathbf{C}^n$ and $\mathbf{C}$, respectively. Descriptions will be given and complemented by various examples for completeness.

1 Introduction

It is a natural goal to describe entire solutions in $\mathbf{C}^n$ to first-order partial differential equations $\frac{\partial w}{\partial z_j} = f(z, w)$, where f is a function in $w \in \mathbf{C}$ and $z \in \mathbf{C}^n$. When $n = 1$, the equations are related to the Malmquist type ordinary differential equations of the form $w' = R(z, w)$, where R is rational in w and $z \in \mathbf{C}$, for which the order of growth of a transcendental meromorphic solution can be described (see [Wi]). It seems hard to describe entire solutions for the equations in the most general form. Here, however, we are interested in partial differential equations of the simple-looking form $\frac{\partial w}{\partial z_j} = p(z) f(w)$, where $f : \mathbf{C} \to \mathbf{P}$ is meromorphic in the complex plane, and p is rational or transcendental in $\mathbf{C}^n$. The special case that $p(z) \equiv 1$ was considered in [LS] for its relations with some other problems, and descriptions of entire solutions of $\frac{\partial w}{\partial z_j} = f(w)$ were given by utilizing an elementary characterization of polynomials. It is known that any entire solutions F of the partial differential equation $\frac{\partial w}{\partial z_j} = f(w)$, where f is a meromorphic function in $\mathbf{C}$, are either of the form $F = a z_j + \alpha$ or of the form $F = a + e^{b z_j + \alpha}$, where a, b are constants and α is an entire function in $z' = (z_1, \ldots, z_{j-1}, z_{j+1}, \ldots, z_n) \in \mathbf{C}^{n-1}$ (see [LS]). However, it is easy to see that this result is no longer true for partial differential equations

$\frac{\partial w}{\partial z_j} = p(z)f(w)$ when p is not a constant. For example, for any entire function $p(z)$ in $\mathbf{C}^n$, the function $F(z) = e^{\int p(z)dz_1}$ is an entire solution of the partial differential equation $\frac{\partial w}{\partial z_1} = p(z)f(w)$ with $f(w) = w$, $w \in \mathbf{C}$. This solution F is clearly not of any of the above forms for general p. It turns out that the results (see Section 2) are quite different when a function $p(z)$, $z \in \mathbf{C}^n$ is involved in the equations; and we will see that whether p is transcendental or not also makes the situations different.

We will state the theorems in the next section. The theorems will be complemented by various examples for the purpose of completeness. The proofs of the results will be given in Section 3.

Unlike in [LS], in proving our results we will employ Nevanlinna theory. We will assume familiarity with basics of the theory, and also basics of one and several complex variables (see, e.g., [BG, Kr, St]).

2 Theorems and examples

We first give the following.

Theorem 1. *Let F be an entire solution in $\mathbf{C}^n$ of the partial differential equation $\frac{\partial w}{\partial z_j} = p(z)f(w)$, say $j = 1$, where f is a meromorphic function in $\mathbf{C}$ and p is a rational function in $\mathbf{C}^n$. Then F must be either a polynomial in $\mathbf{C}^n$ or of one of the following forms:*

(i) $F = \int ap(z)dz_1$;
(ii) $F = a + e^{c \int p(z)dz_1}$;
(iii) $(F - a)^{1-k} = c \int p(z)dz_1$,

where a and $c \neq 0$ are complex numbers, and $k > 1$ is an integer.

Note that in Theorem 1(iii), the solution is an implicit solution, which means that a relation that defines at least one entire solution to the given differential equation.

Remark 1. We give examples to show that each form in Theorem 1 can indeed occur.

(a) Let p be any polynomial in $\mathbf{C}^n$. Set

$$F(z) = \int ap(z)dz_1,$$

where a is a complex number. Then F is an entire solution of the partial differential equation $\frac{\partial w}{\partial z_1} = p(z)f(w)$ with $f(w) = a$, $w \in \mathbf{C}$. This solution is of the form (i) in the theorem.

(b) Let p be any polynomial in $\mathbf{C}^n$. Set

$$F(z) = a + e^{c \int p(z)dz_1},$$

where a and $c \neq 0$ are complex numbers. Then F is an entire solution of the partial differential equation $\frac{\partial w}{\partial z_1} = p(z)f(w)$ with $f(w) = c(w - a)$, $w \in \mathbf{C}$. This solution is of the form (ii) in the theorem.

(c) Let $k > 1$ be an integer, a be a complex number, and $p(z) = \frac{1}{z_1^k}$. Then the relation $(F - a)^{1-k} = (1 - k) \int_1^{z_1} p(z)dz_1 + 1$ is an implicit solution of the partial differential equation $\frac{\partial w}{\partial z_1} = p(z)f(w)$ with $f(w) = (w - a)^k$, $w \in \mathbf{C}$. This solution is of the form (iii) in the theorem. (Note that the entire function $F(z) = a + z_1$ satisfies the relation.)

(d) Let $F(z)$ be a nonzero polynomial of any given degree in $\mathbf{C}^n$. Set $p(z) = \frac{\frac{\partial F}{\partial z_1}}{F(F-1)}$. The function F is an entire solution of the partial differential equation $\frac{\partial w}{\partial z_1} = p(z)f(w)$ with $f(w) = w(w - 1)$, $w \in \mathbf{C}$. This solution is a polynomial of any given degree; and it cannot be written as any of the forms in (i), (ii), (iii) of Theorem 1.

Remark 2. In Theorem 1, we assumed that p is a rational function in $\mathbf{C}^n$. It is a natural question to ask if the theorem holds if p is transcendental in $\mathbf{C}^n$. The answer is negative. Consider the partial differential equation $\frac{\partial w}{\partial z_1} = p(z)f(w)$ with $p(z) = \frac{e^{z_1+z_2+\cdots+z_n}}{e^{e^{z_1+z_2+\cdots+z_n}}}$, an entire function in $\mathbf{C}^n$, and $f(w) = e^w$, $w \in \mathbf{C}$, an entire function in the complex plane. Then the entire function $F = e^{z_1+z_2+\cdots+z_n}$ is a solution of the partial differential equation. It is clear that this solution is not of any form in the theorem.

Nevertheless, we have the following.

Theorem 2. *Let F be an entire solution in $\mathbf{C}^n$ of the partial differential equation $\frac{\partial w}{\partial z_j} = p(z)f(w)$, say $j = 1$, where p and f are entire functions (in $\mathbf{C}^n$ and $\mathbf{C}$, respectively) satisfying that $T(r, f) \neq S(r, p)$. Then, F must be either a polynomial in $\mathbf{C}^n$ or of one of the following forms:*

(i) $F(z) = \int ap(z)dz_1$;

(ii) $F(z) = a + e^{c\int p(z)dz_1}$,

where a and $c \neq 0$ are complex numbers.

On the above, $T(r, f)$ denotes the Nevanlinna Characteristic of f and $S(r, p)$ denotes a quantity that satisfies that $S(r, p) = o\{T(r, p)\}$ for all $r > 0$ outside a set of r of finite Lebesgue measure.

Remark 3. In Theorem 2, we assumed that $T(r, f) \neq S(r, p)$, i.e., f does not grow "more slowly" than p. This condition is inspired by the above example in Remark 2, which satisfies all the conditions of Theorem 2 except the condition that $T(r, f) \neq S(r, p)$. The solution F there is not of any form in Theorem 2. This shows that the condition $T(r, f) \neq S(r, p)$ cannot be dropped in Theorem 2.

Remark 4. We exhibit examples to show that each form in Theorem 2 can indeed occur.

(a) Let $p(z) = 2$ and $F(z) = 4z_1 + \alpha(z_2, \ldots, z_n)$, where α is an entire function in $\mathbf{C}^{n-1}$. The function F is an entire solution of the partial differential equation

$\frac{\partial w}{\partial z_1} = p(z) f(w)$ with $f(w) = 2$, $w \in \mathbf{C}$. Clearly, $T(r, f) \neq o\{T(r, p)\}$. Thus all the conditions of Theorem 2 are satisfied. This solution is of the form (i) in the theorem.

(b) Let p be any polynomial in $\mathbf{C}^n$. Set $F(z) = e^{\int p(z)dz_1}$. Then F is an entire solution of the partial differential equation $\frac{\partial w}{\partial z_1} = p(z) f(w)$ with $f(w) = w$, $w \in \mathbf{C}$. Clearly, $T(r, f) \neq o\{T(r, p)\}$. Thus, all the conditions of Theorem 2 are satisfied. This solution is of the form (ii) in the theorem.

(c) Let $F(z) = z_1 + z_2 + \cdots + z_n$ for $z = (z_1, z_2, \ldots, z_n)$ in $\mathbf{C}^n$. Set $p(z) = \frac{1}{e^{z_1+z_2+\cdots+z_n}}$, which is entire. The function F is an entire solution of the partial differential equation $\frac{\partial w}{\partial z_1} = p(z) f(w)$ with $f(w) = e^w$, $w \in \mathbf{C}$. Clearly, $T(r, f) \neq o\{T(r, p)\}$. Thus all the conditions of Theorem 2 are satisfied. This solution is a polynomial; and it cannot be written as any of the forms (i) and (ii) in Theorem 2.

We refer to [L] for some related results on entire solutions to certain nonlinear partial differential equations.

3 Proofs of the theorems

Proof of Theorem 1. Let F be an entire solution in $\mathbf{C}^n$ of the partial differential equation $\frac{\partial w}{\partial z_1} = p(z) f(w)$, i.e.,

$$\frac{\partial F(z)}{\partial z_1} = p(z) f(F(z)), \quad z \in \mathbf{C}^n. \tag{1}$$

If F is a polynomial in $\mathbf{C}^n$, the conclusion of the theorem is already true. Thus we assume that F is not a polynomial in the following proof. We can also assume in the following that $p \not\equiv 0$ and $f \not\equiv 0$, since otherwise $\frac{\partial F(z)}{\partial z_1} \equiv 0$ and thus F has the form (i) in Theorem 1 with $a = 0$.

We assert that f has at most one pole in $\mathbf{C}$. Suppose to the contrary that f has two distinct poles w_1 and w_2. By (1), the zeros of $F - w_1$ and $F - w_2$ must be zeros of p, since otherwise the left-hand side, which is an entire function, would have poles, which is absurd. By the Nevanlinna Second Fundamental Theorem, we then have that

$$T(r, F) \leq N\left(r, \frac{1}{F - w_1}\right) + N\left(r, \frac{1}{F - w_2}\right) + S(r, F)$$

$$\leq 2N\left(r, \frac{1}{p}\right) + S(r, F) = O\{\log r\} + S(r, F)$$

which implies that F is a polynomial, a contradiction.

We next discuss two cases: f has no poles and f has one pole.

Case (i). f has one pole, say, w_0. Then we can write f into the following form: $f(w) = \frac{1}{(w-w_0)^m} g(w)$, $w \in \mathbf{C}$, where $m > 0$ is an integer, and g is an entire function in $\mathbf{C}$ with $g(w_0) \neq 0$. By (1), a zero of $F(z) = w_0$ must be a zero of p. Thus

$F(z) = w_0 + p_1(z)e^{q(z)}$, $z \in \mathbf{C}^n$, where p_1 is a nonconstant polynomial in $\mathbf{C}^n$, and q is a nonconstant entire function in $\mathbf{C}^n$, in view of the fact that F is not a polynomial. By (1) again, we obtain that

$$e^{q(z)}\left(p_1(z)\frac{\partial q(z)}{\partial z_1} + \frac{\partial p_1(z)}{\partial z_1}\right) = p(z)\frac{g(w_0 + p_1(z)e^{q(z)})}{(p_1(z)e^{q(z)})^m},$$

or

$$\frac{1}{p(z)}\left(\frac{\partial q(z)}{\partial z_1} + \frac{\frac{\partial p_1(z)}{\partial z_1}}{p_1(z)}\right) = \frac{g(w_0 + p_1(z)e^{q(z)})}{(p_1(z)e^{q(z)})^{m+1}} := h(p_1(z)e^{q(z)}), \qquad (2)$$

where $h(w) = \frac{g(w_0+w)}{w^{m+1}}$, $w \in \mathbf{C}$, is an entire function in the complex plane.

We assert that h is a constant. Suppose to the contrary that h is not a constant. Then we can show that q must be a constant. To see this, recall the following theorem in [CLY]: If F is a transcendental meromorphic function in $\mathbf{C}$ and G is a transcendental entire function in $\mathbf{C}^n$, then $\lim_{r\to\infty} \frac{T(r,F(G))}{T(r,G)} = +\infty$. Thus, if q is not a constant, we would have that $T(r, q) = o\{T(r, e^q)\}$ when q is transcendental. This is clearly also true when q is a nonconstant polynomial. On the other hand, we have that $T(r, \frac{\partial h(z)}{\partial z_1}) = O\{T(r, h)\}$ for any meromorphic function h in $\mathbf{C}^n$ outside a set of r of finite Lebesgue measure (see, e.g., [St, Vi]). Thus we would have from (2) that

$$T(r, q) = o\{T(r, e^q)\} = o\{T(r, p_1 e^q)\}$$
$$= o\{T(r, h(p_1 e^q))\}$$
$$= o\left\{T\left(r, \frac{1}{p(z)}\left(\frac{\partial q(z)}{\partial z_1} + \frac{\frac{\partial p_1(z)}{\partial z_1}}{p_1(z)}\right)\right)\right\}$$
$$= o\{T(r, p)\} + o\{T(r, q)\} + o\{T(r, p_1)\} = o\{T(r, q)\} + o\{\log r\}$$

outside a set of r of finite Lebesgue measure. Thus $T(r, q) = o\{\log r\}$ outside a set of r of finite Lebesgue measure, which implies that q is a constant, a contradiction. This shows that q is a constant. Hence $F(z) = w_0 + p_1(z)e^{q(z)}$ is a polynomial, which contradicts the fact that F is not a polynomial assumed in the beginning of the proof.

Therefore, h is a constant. That is, $\frac{g(w_0+w)}{w^{m+1}} = c$, a constant. But then we have that $f(w_0 + w) = \frac{1}{w^m}g(w_0 + w) = cw$ for $w \neq 0$, or $f(w) = c(w - w_0)$ for all $w \neq w_0$. This implies that w_0 is a removable singularity of f, a contradiction to the assumption that w_0 is a pole of f. (Thus Case (i) actually does not occur.)

Case (ii). f has no poles in the complex plane. In this case, f is an entire function in the complex plane. It is easy to see that f must be a polynomial, since otherwise we would have from (1) that

$$T(r, F) = o\{T(r, f(F))\} = o\left\{T\left(r, \frac{\partial F}{\partial z_1}\right) + T(r, p)\right\} + O(1)$$
$$= o\{T(r, F)\} + o\{\log r\}$$

outside a set of r of finite Lebesgue measure, which implies that F is a constant, again a contradiction to the fact that F is not a polynomial assumed in the beginning of the proof. We can now write $f(w) = C\Pi_{j=1}^{l}(w - a_j)^{k_j}$, where C is a nonzero constant, a_js are l distinct complex numbers, and k_js are positive integers. We may write (1) into

$$\frac{\partial F(z)}{\partial z_1} = Cp(z)\Pi_{j=1}^{l}(F(z) - a_j)^{k_j}. \tag{3}$$

We next show that $l \leq 1$ in (3). To this end, we write $F(z)$ to a Taylor series in z_1 with coefficients being functions of $z_2, z_3, \ldots, z_n$. This series is either a polynomial in z_1 or transcendental in z_1 (with functional coefficients in $z_2, z_3, \ldots, z_n$). If the series is a polynomial in z_1, then by comparing the leading coefficients of z_1 in both sides of (3), we see that if $l > 1$ then the leading coefficient of F is not transcendental and the zero set of $F - a_j$ must be a subset of the algebraic variety consisting of the poles of p and the zeros of the leading coefficient of F. By the Nevanlinna Second Fundamental Theorem, we know that the transcendental function F has at most one such "Picard value" a_j. This shows that $l \leq 1$ in this case. Now, if the series is transcendental in z_1, we then can take complex numbers $z_2^*, z_3^*, \ldots, z_n^*$ such that $F(z_1, z_2^*, z_3^*, \ldots, z_n^*)$ is a transcendental entire function in z_1. By (3), we have that

$$\frac{\partial F(z_1, z_2^*, \ldots, z_n^*)}{\partial z_1} = Cp(z_1, z_2^*, \ldots, z_n^*)\Pi_{j=1}^{l}(F(z_1, z_2^*, \ldots, z_n^*) - a_j)^{k_j}.$$

We see from this identity that for each a_j, the zero set of $F(z_1, z_2^*, \ldots, z_n^*) - a_j$ in the complex plane must be a subset of the set consisting of the poles of $p(z_1, z_2^*, \ldots, z_n^*)$. By the Nevanlinna Second Fundamental Theorem again, there exists at most one such a_j. That is, $l \leq 1$. Thus, in any case, the equality (3) is reduced to

$$\frac{\partial F(z)}{\partial z_1} = Cp(z)(F(z) - a)^k,$$

where a is a complex number and k is a nonnegative integer. If $k > 1$, by solving the above partial differential equation we deduce that

$$(F - a)^{1-k} = (1 - k) \int Cp(z)dz_1.$$

This solution is of the form (iii) in the theorem. If $k = 1$, by solving the above equation we obtain that $F(z) = a + e^{c \int p(z)dz_1}$, which is of the form (ii) in the theorem. If $k = 0$, by solving the above equation we obtain that $F(z) = \int Cp(z)dz_1$, which is of the form (i) in the theorem. The proof is thus complete. $\square$

Proof of Theorem 2. Let F be an entire solution in $\mathbf{C}^n$ of the partial differential equation $\frac{\partial w}{\partial z_1} = f(w)$, i.e.,

$$\frac{\partial F(z)}{\partial z_1} = p(z)f(F(z)), \quad z \in \mathbf{C}^n. \tag{4}$$

If F is a polynomial in $\mathbf{C}^n$, the conclusion of the theorem is already true. Thus we assume that F is not a polynomial in the following proof.

We assert that f must be a polynomial. Suppose to the contrary that f is transcendental. Then it follows from (4) that

$$T(r, f(F)) \leq T\left(r, \frac{\partial F(z)}{\partial z_1}\right) + T(r, p) + O(1)$$
$$\leq T(r, F) + T(r, p) + S(r, F)$$
$$\leq o\{T(r, f(F))\} + T(r, p)$$

outside a set of r of finite Lebesgue measure, using the result in [CLY] we mentioned in the proof of Theorem 1. We thus have that

$$T(r, f(F)) = O\{T(r, p)\} \tag{5}$$

outside a set of r of finite Lebesgue measure. Next, we use another result in [CLY]: If f is a transcendental entire (not meromorphic) function in $\mathbf{C}$ and g is a transcendental entire function in $\mathbf{C}^n$, then $\lim_{r \to \infty} \frac{T(r, f(g))}{T(r, f)} = +\infty$. (Note that this result is false when f is not entire.) Apply this result to our situation, we then have, in view of (5), that

$$T(r, f) = o\{T(r, f(F))\} = o\{T(r, p)\}$$

outside a set of r of finite Lebesgue measure, i.e., $T(r, f) = S(r, p)$, which contradicts the assumption of the theorem.

Thus f is a polynomial. If $\frac{\partial F}{\partial z_1} \equiv 0$, then $F = \alpha(z_2, z_3, \ldots, z_n)$ for an entire function α in $\mathbf{C}^{n-1}$, which has the form (i) in the theorem with $a = 0$. Hence, we assume that $\frac{\partial F}{\partial z_1} \not\equiv 0$ in the following. Then we may choose complex numbers $w_2, w_3, \ldots, w_n$ such that $F(z_1, w_2, w_3, \ldots, w_n)$ is not a constant function in z_1. By (4), we have that

$$\frac{\partial F(z_1, w_2, \ldots, w_n)}{\partial z_1} = Cp(z_1, w_2, \ldots, w_n)\Pi^l_{j=1}(F(z_1, w_2, \ldots, w_n) - a_j)^{k_j}, \tag{6}$$

where C is a nonzero constant, a_js are $l \geq 0$ distinct complex numbers, and k_js are positive integers. By comparing the multiplicities at a possible zero of $F - a_j$ for both sides of (6), we see that $F - a_j$ cannot have any zeros. By the Picard Theorem, we have at most one such a_j. Thus the equality (4) is reduced to

$$\frac{\partial F(z)}{\partial z_1} = Cp(z)(F(z) - a)^k, \tag{7}$$

where a is a complex number and k is a nonnegative integer. Recall that f is a polynomial. This actually also implies that p is a polynomial, since otherwise we would have that $\frac{T(r,p)}{\log r} \to \infty$ and thus that $T(r, f) = O\{\log r\} = o\{T(r, p)\}$, which contradicts the condition of the theorem that $T(r, f) \neq o\{T(r, p)\}$. Note that $F - a = e^g$ for an entire function g in z_1, where $F := F(z_1, w_2, \ldots, w_n)$. We then have by (7) that

$$\frac{\partial g}{\partial z_1} = Cpe^{(k-1)g},$$

which implies that, if $k > 1$,

$$T(r, g) = o\{T(r, e^{(k-1)g})\}$$

$$= o\left\{T\left(r, \frac{1}{Cp}\frac{\partial g}{\partial z_1}\right)\right\} = o\{T(r, g)\} + o\{\log r\}.$$

Thus $T(r, g) = o\{\log r\}$, which implies that g is a constant and so that F is a constant, a contradiction.

Thus we have showed that $k = 0$ or $k = 1$. If $k = 0$, by solving the above equation we obtain that $F(z) = \int Cp(z)dz_1$, which is of the form (i) in the theorem. If $k = 1$, by solving the equation we obtain that $F(z) = a + e^{c\int p(z)dz_1}$, which is of the form (ii) of the theorem. This completes the proof of Theorem 2.   □

Acknowledgments The authors are supported in part by NSF grants.

References

[BG]   C. A. Berenstein and R. Gay, *Complex Variables*, Springer-Verlag, New York, 1991.

[CLY]   D. C. Chang, B. Q. Li, and C. C. Yang, On composition of meromorphic functions in several complex variables, *Forum Math.*, **7** (1995), 77–94.

[Kr]   S. Krantz, *Function Theory of Several Complex Variables*, Wiley, New York, 1982.

[L]   B. Q. Li, Entire solutions of certain partial differential equations in $\mathbf{C}^n$, *Israel J. Math.*, to appear.

[LS]   B. Q. Li and E. G. Saleeby, Entire solutions of first-order partial differential equations, *Complex Variables*, **48** (2003), 657–661.

[St]   W. Stoll, *Introduction to the Value Distribution Theory of Meromorphic Functions*, Springer-Verlag, New York, 1982.

[Vi]   A. Vitter, The lemma of the logarithmic derivative in several complex variables, *Duke Math. J.*, **44** (1977), 89–104.

[Wi]   H. Wittich, Eindeutige Lösungen der Differentialgleichung $w' = R(z, w)$, *Math. Z.*, **74** (1960), 278–288.

Hermite Operator on the Heisenberg Group*

Ovidiu Calin[1], Der-Chen Chang[2], and Jingzhi Tie[3]

[1] Department of Mathematics
Eastern Michigan University
Ypsilanti, MI 48197
USA
ocalin@emunix.emich.edu
[2] Department of Mathematics
Georgetown University
Washington, DC 20057
USA
chang@math.georgetown.edu
[3] Department of Mathematics
University of Georgia
Athens, GA 30602-7403
USA
jtie@math.uga.edu

Dedicated to Professor Carlos Berenstein on his 60th birthday.

Summary. In this article, we first introduce a new geometric method based on multipliers to compute heat kernels for operators with potentials. Using the heat kernel, we may compute the fundamental solution for the Hermite operator with a singularity at an arbitrary point on the Heisenberg group. As a consequence, one may obtain the fundamental solutions for the sub-Laplacian in $\mathbf{H}_n$ and the Grusin operator in $\mathbb{R}^n$.

1 Introduction

The Heisenberg group and its sub-Laplacian are at the crossroads of many domains of analysis and geometry: nilpotent Lie group theory, hypoelliptic second order partial differential equations, strongly pseudoconvex domains in complex analysis, probability theory of degenerate diffusion process, sub-Riemannian geometry, control theory and semiclassical analysis of quantum mechanics, see, e.g., [BGGr, BGr, CCGr, CGr, CT1, CT2].

The Heisenberg group is the simplest nilpotent Lie group with underlying manifold $\mathbb{R}^{2n+1}$ and multiplicative law

* The second author is partially supported by a William Fulbright Research grant and a competitive research grant at Georgetown University.

$$[\mathbf{x}, t] \cdot [\mathbf{y}, s] = \left[\mathbf{x} + \mathbf{y}, t + s + 2 \sum_{j=1}^{n} a_j (x_{j+n} y_j - x_j y_{j+n}) \right].$$

The nonisotropic dilation $[x_1, \ldots, x_{2n}, t] \mapsto [\delta x_1, \ldots, \delta x_{2n}, \delta^2 t]$ for $\delta > 0$ defines an automorphism on the group $\mathbf{H}_n$.

The Heisenberg Lie algebra is the vector space of left invariant vector fields with the usual Lie bracket

$$[\mathbf{X}, \mathbf{Y}] = \mathbf{X}\mathbf{Y} - \mathbf{Y}\mathbf{X}$$

with the following basis:

$$\mathbf{X}_j = \frac{\partial}{\partial x_j} - 2a_j x_{j+n} \frac{\partial}{\partial t}, \qquad \mathbf{X}_{j+n} = \frac{\partial}{\partial x_{j+n}} + 2a_j x_j \frac{\partial}{\partial t}, \qquad j = 1, \ldots, n,$$

and $\mathbf{T} = \frac{\partial}{\partial t}$. Here $a_j = a_{j+n}$ for $j = 1, \ldots, n$. It is easy to see that $\mathbf{X}_1, \ldots, \mathbf{X}_{2n}$ satisfy the Heisenberg uncertainty principle:

$$[\mathbf{X}_j, \mathbf{X}_{j+n}] = -4a_j \frac{\partial}{\partial t} \quad \text{and} \quad [\mathbf{X}_j, \mathbf{X}_k] = 0, \quad j \neq k, \quad j, k = 1, \ldots, n.$$

Hence $\mathbf{H}_n$ is a noncommutative Lie group of step 2, i.e., $\mathbf{X}_1, \ldots, \mathbf{X}_{2n}$ and their first brackets yield the tangent bundle $T\mathbf{H}_n$. The sub-Laplacian on $\mathbf{H}_n$ is defined by

$$\mathcal{L}_\lambda = i\lambda \frac{\partial}{\partial t} - \sum_{j=1}^{n} (\mathbf{X}_j^2 + \mathbf{X}_{j+n}^2)$$

$$= i\lambda \frac{\partial}{\partial t} - \sum_{j=1}^{2n} \left[\frac{\partial^2}{\partial x_j^2} + 4a_j^2 x_j^2 \frac{\partial^2}{\partial t^2} \right] + 2 \sum_{j=1}^{n} a_j \left(x_j \frac{\partial}{\partial x_{j+n}} - x_{j+n} \frac{\partial}{\partial x_j} \right) \frac{\partial}{\partial t}.$$

This operator is a sum of squares of $2n$ "horizontal" vector fields, and it is therefore not elliptic, although it is hypoelliptic (see Hörmander [H1]), i.e., the solution u of $\mathcal{L}_\lambda u = f$ is smooth whenever $f \in C^\infty(\mathbf{H}_n)$ provided $\lambda \neq \sum_{j=1}^{n} (2k_j + 1)a_j$, where $(k_1, \ldots, k_n) \in (\mathbb{Z}_+)^n$ (see [BCT] and [CT1]). It is easy to see that the vector fields $\mathbf{X}_1, \ldots, \mathbf{X}_{2n}$ and $\mathbf{T} = \frac{\partial}{\partial t}$ and the operator $\mathcal{L}_\lambda$ are left-invariant with respect to the Heisenberg translation. The operator $\mathcal{L}_\alpha$ arises naturally when the operator $\square_b = \bar{\partial}_b \bar{\partial}_b^* + \bar{\partial}_b^* \bar{\partial}_b$ acts on $(0, q)$-forms on the group $\mathbf{H}_n$. Here $\bar{\partial}_b$ is the tangential Cauchy–Riemann operator and $\bar{\partial}_b^*$ is the Hilbert space adjoint of $\bar{\partial}_b$. Moreover, the group $\mathbf{H}_n$ can be identified as the boundary of the Siegel upper space:

$$\Omega_{n+1} = \left\{ (z_1, \ldots, z_{n+1}) \in \mathbb{C}^{n+1} : \text{Im}(z_{n+1}) > \sum_{j=1}^{n} a_j |z_j|^2 \right\},$$

which plays a very important role in analysis on strongly pseudoconvex domains.

Here we are interested in the fundamental solution of the Hermite operator

$$\mathcal{H}_\alpha = \alpha + \sum_{j=1}^{2n}(\beta_j^2 x_j^2 - X_j^2) + i\lambda \mathbf{T} = \mathcal{L}_\lambda + \alpha + \sum_{j=1}^{2n}\beta_j^2 x_j^2$$

in $\mathbf{H}^n$. More precisely, we are looking for a distribution $K_\alpha(\mathbf{x}, \mathbf{y})$ such that

$$\left[\alpha + \sum_{j=1}^{2n}\left(\beta_j^2 x_j^2 - X_j^2\right) + i\lambda \mathbf{T}\right] K_\alpha(\mathbf{x}, \mathbf{y}) = \delta(\mathbf{x} - \mathbf{y}). \tag{1}$$

Since the operator $\mathcal{H}_\alpha$ is not invariant under the Euclidean group action or the Heisenberg group action, we have to compute the fundamental solution with singularity at any point $\mathbf{y}$. One way to achieve this goal is to take the Fourier transform with respect to the t-variable (which is the center of the group) and reduce the operator as a Hermite operator in $\mathbb{R}^{2n}$ with parameters α, λ and the dual variable τ of t. Using the results for Hermite operator (see, e.g., [B] and [CT3]) and inverse Fourier transform, one may solve the problem.

Here we shall use the Hamilton–Jacobi theory to find the heat kernel for the heat equation with potentials. The idea is to write down the Euler–Lagrange system of equations for the Lagrangian and to characterize the system qualitatively from the conservation law point of view. In general, these systems cannot be solved explicitly. For simple equations, one may characterize the solutions by finding the first integrals of motions. We will demonstrate this method by solving the operator with potential $U(x) = \beta^2 x^2$. Then we shall give some examples and comments for general potentials at the end of Section 2. For more details, readers can consult the book [CC].

2 Heat kernel for the Hermite operator in $\mathbb{R}^2$

We start with the Hermite operator

$$H = \frac{1}{2}\left(\frac{d^2}{dx^2} - \lambda^2 x^2\right),$$

where $\lambda \in \mathbb{R}_+$ is a nonnegative real parameter. We associate the Hamiltonian function as half of the principal symbol

$$H(\xi, x) = \frac{1}{2}(\xi^2 - \lambda^2 x^2). \tag{2}$$

The Hamiltonian system is

$$\dot{x} = H_\xi = \xi \quad \text{and} \quad \dot{\xi} = -H_x = \lambda^2 x.$$

As we are interested in finding the geodesic between the points $x_0, x \in \mathbb{R}$, $x(s)$ will satisfy the boundary problem

$$\ddot{x} = \lambda^2 x \quad \text{with } x(0) = x_0, x(t) = x.$$

The conservation of energy law is

$$\frac{1}{2}\dot{x}^2(s) - \frac{1}{2}\lambda^2 x^2(s) = E,$$

where E is the energy constant. This can be used to obtain an ODE for the solution $x(s)$:

$$\frac{dx}{ds} = \sqrt{2E + \lambda^2 x^2} \implies \frac{dx}{\sqrt{2E + \lambda^2 x^2}} = ds.$$

Integrating between $s = 0$ and $s = t$, with $x(0) = x_0$ and $x(t) = x$, yields

$$\int_{x_0}^{x} \frac{du}{\sqrt{2E + \lambda^2 u^2}} = t \iff \int_{v_0}^{v} \frac{dv}{\sqrt{1 + v^2}} = \lambda t,$$

with $v = \frac{\lambda x}{\sqrt{2E}}$ and $v_0 = \frac{\lambda x_0}{\sqrt{2E}}$. Integrating yields

$$\sinh^{-1}(v) - \sinh^{-1}(v_0) = \lambda t \iff \sinh^{-1}(v) = \sinh^{-1}(v_0) + \lambda t,$$

which is equivalent to

$$v = \sinh(\sinh^{-1}(v_0) + \lambda t)$$
$$\iff v = v_0 \cosh(\lambda t) + \cosh(\sinh^{-1}(v_0)) \sinh(\lambda t)$$
$$\iff v = v_0 \cosh(\lambda t) + \sqrt{1 + v_0^2} \sinh(\lambda t).$$

Hence

$$\frac{\lambda x}{\sqrt{2E}} = \frac{\lambda x_0}{\sqrt{2E}} \cosh(\lambda t) + \sqrt{1 + \frac{\lambda^2 x_0^2}{2E}} \sinh(\lambda t)$$

and

$$\lambda x = \lambda x_0 \cosh(\lambda t) + \sqrt{2E + \lambda^2 x_0^2} \sinh(\lambda t).$$

It follows that

$$\frac{\lambda(x - x_0 \cosh(\lambda t))}{\sinh(\lambda t)} = \sqrt{2E + \lambda^2 x_0^2}.$$

Solving for E yields

$$2E = \frac{\lambda^2 (x - x_0 \cosh(\lambda t))^2}{\sinh(\lambda t)^2} - \lambda^2 x_0^2$$
$$= \frac{\lambda^2 (x^2 - 2xx_0 \cosh(\lambda t) + x_0^2 \cosh(\lambda t)^2 - x_0^2 \sinh(\lambda t)^2)}{\sinh(\lambda t)^2}$$
$$= \frac{\lambda^2 (x^2 + x_0^2 - 2xx_0 \cosh(\lambda t))}{\sinh(at)^2}.$$

Proposition 1. *The energy along a geodesic derived from the Hamiltonian* (2) *between the points x_0 and x is*

$$E = \frac{\lambda^2(x^2 + x_0^2 - 2xx_0\cosh(\lambda t))}{2\sinh(at)^2}. \tag{3}$$

Making $x_0 = 0$, we obtain the following result.

Corollary 1. *The energy along a geodesic derived from the Hamiltonian* (2) *joining the origin and x is given by*

$$E = \frac{\lambda^2 x^2}{2\sinh(\lambda t)^2}. \tag{4}$$

We note that if we take the limit $a \to 0$ in (3), we obtain the Euclidean energy

$$\lim_{\lambda\to 0} E = \lim_{\lambda\to 0} \frac{\lambda^2 t^2}{\sinh(\lambda t)^2} \frac{(x^2 + x_0^2 - 2xx_0\cosh(\lambda t))}{2t^2} = \frac{(x-x_0)^2}{2t^2}.$$

2.1 The action function

Let $S = S(x_0, x, t)$ be the action with initial point x_0 and final point x, within time t. The action satisfies the Hamilton–Jacobi equation

$$\partial_t S + H(\nabla S) = 0.$$

One may note that

$$H(\xi, x) = \frac{1}{2}(\xi^2 - \lambda^2 x^2) = \frac{1}{2}\dot{x}^2 - \frac{1}{2}\lambda^2 x^2 = E,$$

and hence $\partial_t S = -E$. Using (3) yields

$$\begin{aligned}
\frac{\partial S}{\partial t} &= -\frac{\lambda^2(x^2 + x_0^2 - 2xx_0\cosh(\lambda t))}{2\sinh(\lambda t)^2} \\
&= \frac{\lambda}{2}(x^2 + x_0^2)\frac{\partial}{\partial t}\coth(\lambda t) - \lambda xx_0\frac{\partial}{\partial t}\frac{1}{\sinh(\lambda t)} \\
&= \frac{\partial}{\partial t}\left[\frac{\lambda}{2}(x^2 + x_0^2)\coth(\lambda t) - \frac{\lambda xx_0}{\sinh(\lambda t)}\right].
\end{aligned}$$

Hence we arrived at the action

$$\begin{aligned}
S(x_0, x, t) &= \frac{a}{2}\left[(x^2 + x_0^2)\coth(\lambda t) - \frac{2xx_0}{\sinh(\lambda t)}\right] \\
&= \frac{\lambda}{2}\frac{1}{\sinh(\lambda t)}[(x^2 + x_0^2)\cosh(\lambda t) - 2xx_0]. \tag{5}
\end{aligned}$$

One may note easily that

$$\lim_{\lambda\to 0} S = \frac{(x-x_0)^2}{2t},$$

which is the Euclidean action.

Lemma 1. *We have*

(i) $(\partial_x S)^2 = \lambda^2 x^2 + 2E$,

(ii) $\partial_x^2 S = \lambda \coth(\lambda t)$.

Proof.

(i) Differentiating in (5) with respect to the x-variable yields

$$\partial_x S = \frac{\lambda}{\sinh(\lambda t)}(x \cosh(\lambda t) - x_0), \tag{6}$$

$$
\begin{aligned}
(\partial_x S)^2 &= \frac{\lambda^2 (x^2 \cosh^2(\lambda t) + x_0^2 - 2xx_0 \cosh(\lambda t))}{\sinh^2(\lambda t)} \\
&= \frac{\lambda^2 (x^2 + x^2 \sinh^2(\lambda t) + x_0^2 - 2xx_0 \cosh(\lambda t))}{\sinh^2(\lambda t)} \\
&= a^2 x^2 + \frac{\lambda^2 (x^2 + x_0^2 - 2xx_0 \cosh(\lambda t))}{\sinh^2(\lambda t)} \\
&= \lambda^2 x^2 + 2E.
\end{aligned}
$$

(ii) Again, differentiating in (6) with respect to the x-variable yields

$$\partial_x^2 S = \frac{\lambda}{\sinh(\lambda t)} \cosh(\lambda t) = \lambda \coth(\lambda t). \qquad \square$$

We shall look for a fundamental solution of the heat equation which has a representation as follows.

$$K(x_0, x, t) = V(t)e^{kS(x_0, x, t)}, \tag{7}$$

where $V(t)$ will satisfy a volume function equation and k is a real constant. Lemma 1 provides

$$
\begin{aligned}
\partial_t K &= V'(t)e^{kS} + V(t)ke^{kS}\partial_t S \\
&= e^{kS}(V'(t) - kV(t)E).
\end{aligned}
$$

Hence

$$
\begin{aligned}
\partial_x e^{kS} &= ke^{kS}\partial_x S, \\
\partial_x^2 e^{kS} &= k^2 e^{kS}(\partial_x S)^2 + ke^{kS}\partial_x^2 S \\
&= ke^{kS}[k(\partial_x g)^2 + \partial_x^2 S] \\
&= ke^{kS}[k(\lambda^2 x^2 + 2E) + \lambda \coth(\lambda t)].
\end{aligned}
$$

We shall find the heat kernel using a multiplier method. Let

$$P = \partial_t - \partial_x^2 + \alpha \lambda^2 x^2, \tag{8}$$

where α is a real multiplier, which will be determined such that $PK(x_0, x, t) = 0$ for any $t > 0$:

$$PK(x_0, x, t) = e^{kS}(V'(t) - kEV(t))$$
$$- ke^{kS}(k(\lambda^2 x^2 + 2E) + \lambda \coth(\lambda t))V(t)\alpha\lambda^2 x^2 e^{kS}V(t)$$
$$= e^{kS}V(t)\left[\frac{V'(t)}{V(t)} - kE - k^2(\lambda^2 x^2 + 2E) - k\lambda \coth(\lambda t) + \alpha\lambda^2 x^2\right]$$
$$= e^{kS}V(t)\left[\frac{V'(t)}{V(t)} - kE - k^2\lambda^2 x^2 - 2k^2 E + \alpha\lambda^2 x^2 - ka \coth(\lambda t)\right]$$
$$= e^{kS}V(t)\left[\frac{V'(t)}{V(t)} - kE(2k+1) + (\alpha - k^2)\lambda^2 x^2 - ka \coth(\lambda t)\right].$$

In order to eliminate the middle two terms in the brackets, we choose $k = -\frac{1}{2}$ and $\alpha = \frac{1}{4}$. Let $\beta = \frac{\lambda}{2} > 0$. Then the operator (8) becomes

$$P = \partial_t - \partial_x^2 + \beta^2 x^2 \tag{9}$$

and

$$PK(x_0, x, t) = K(x_0, x, t)\left(\frac{V'(t)}{V(t)} + \beta \coth(2\beta t)\right).$$

We shall choose $V(t)$ such that

$$\frac{V'(t)}{V(t)} = -\beta \coth(2\beta t), \quad t > 0.$$

Integrating yields

$$\ln V(t) = -\frac{1}{2}\ln(\sinh(2\beta t)) \implies V(t) = \frac{C}{\sqrt{\sinh(2\beta t)}}.$$

Using the action (5), the fundamental solution formula (7) becomes

$$K(x_0, x, t) = \frac{C}{\sqrt{\sinh(2\beta t)}}e^{-\frac{2\beta}{4}\frac{1}{\sinh(2\beta t)}[(x^2 + x_0^2)\cosh(2\beta t) - 2xx_0]}$$
$$= \frac{C}{\sqrt{2\beta t}}\sqrt{\frac{2\beta t}{\sinh(2\beta t)}}e^{-\frac{1}{4t}\cdot\frac{2\beta t}{\sinh(2\beta t)}[(x^2 + x_0^2)\cosh(2\beta t) - 2xx_0]}.$$

We shall find the constant C investigating the limit case $\beta \to 0$, when the operator (9) becomes the usual one-dimensional heat operator $\partial_t - \partial_x^2$. As $\frac{2\beta t}{\sinh(2\beta t)} \to 1$, the above fundamental solution becomes

$$K(x_0, x, t) \sim \frac{C}{\sqrt{2\beta t}}e^{\frac{1}{4t}(x - x_0)^2}, \quad \beta \to 0.$$

By comparison with the fundamental solution for the usual heat operator, which is

$$\frac{1}{\sqrt{4\pi t}}e^{\frac{1}{4t}(x - x_0)^2},$$

we find $C = \sqrt{\frac{\beta}{2\pi}}$. We arrive at the following result.

Theorem 1. *Let $\beta \geq 0$. The fundamental solution for the operator $P = \partial_t - \partial_x^2 + \beta^2 x^2$ is*

$$
K(x_0, x, t) = \frac{1}{\sqrt{4\pi t}} \sqrt{\frac{2\beta t}{\sinh(2\beta t)}} e^{-\frac{1}{4t}\frac{2\beta t}{\sinh(2\beta t)}[(x^2+x_0^2)\cosh(2\beta t)-2xx_0]}
$$

$$
= \frac{1}{\sqrt{2\pi}} \sqrt{\frac{\beta}{\sinh(2\beta t)}} e^{-\frac{\beta(x^2+x_0^2)\cosh(2\beta t)-2\beta xx_0}{2\sinh(2\beta t)}}, \quad t > 0.
$$

The computations are similar in the case when $\beta = -i\gamma$. Using $\cosh(i\gamma t) = \cos(\gamma t)$ and $\sinh(2i\gamma t) = i \sin(2\gamma t)$, we obtain a dual theorem.

Theorem 2. *Let $\gamma \geq 0$. The fundamental solution for the operator $P = \partial_t - \partial_x^2 - \gamma^2 x^2$ is*

$$
K(x_0, x, t) = \frac{1}{\sqrt{4\pi t}} \sqrt{\frac{2\gamma t}{\sin(2\gamma t)}} e^{-\frac{1}{4t}\frac{2\gamma t}{\sin(2\gamma t)}[(x^2+x_0^2)\cos(2\gamma t)-2xx_0]}
$$

$$
= \frac{1}{\sqrt{2\pi}} \sqrt{\frac{\gamma}{\sin(2\gamma t)}} e^{-\frac{\gamma(x^2+x_0^2)\cos(2\gamma t)-2\gamma xx_0}{2\sin(2\gamma t)}}, \quad t > 0.
$$

2.2 The harmonic oscillator $\partial_t - \sum_{j=1}^n (\partial_{x_j}^2 \pm \lambda_j^2 x_j^2)$

Consider the Hermite operator in $\mathbb{R}^n$

$$
\frac{1}{2}\left(\Delta_n - \sum_{j=1}^n \lambda_j^2 x_j^2\right) = \frac{1}{2}\sum_{j=1}^n (\partial_{x_j}^2 - \lambda^2 x_j^2),
$$

where $\lambda_j \geq 0$ for $j = 1, \ldots, n$. The associated Hamiltonian is

$$
H(x_1, \ldots, x_n, \xi_1, \ldots, \xi_n) = \sum_{j=1}^n \frac{1}{2}(\xi_j^2 - \lambda_j^2 x_j^2),
$$

with the Hamiltonian system

$$
\dot{x}_j = H_{\xi_j} = \xi_j \quad \text{and} \quad \dot{\xi}_j = -H_{x_j} = \lambda_j^2 x_j, \quad j = 1, \ldots, n.
$$

The geodesic $\mathbf{x}(s)$ starting at $\mathbf{x}_0 = (x_1^0, \ldots x_n^0)$ and having the final point $\mathbf{x} = (x_1, \ldots, x_n)$ satisfies the equations

$$
\ddot{x}_j = \lambda_j^2 x_j \quad \text{with } x_j(0) = x_j^0, \, x_j(t) = x_j, \quad j = 1 \ldots n.
$$

As in the one-dimensional case, we have the law of conservation of energy

$$
\dot{x}_j^2(s) - \lambda_j^2 x_j^2(s) = 2E_j, \quad j = 1, \ldots, n,
$$

where E_j is the energy constant for the jth component. The total energy, which is the Hamiltonian, is given by

$$H = \frac{1}{2} \sum_{j=1}^{n} (\dot{x}_j^2 - \lambda_j^2 x_j^2) = E_1 + \cdots + E_n = E \text{(constant)}.$$

Proposition 1 yields

$$E_j = \frac{\lambda_j^2 [x_j^2 + (x_j^0)^2 - 2x_j x_j^0 \cosh(\lambda_j t)]}{2 \sinh^2(\lambda_j t)},$$

and hence

$$H = E = \sum_{j=1}^{n} E_j = \sum_{j=1}^{n} \frac{\lambda_j^2 [x_j^2 + (x_j^0)^2 - 2x_j x_j^0 \cosh(\lambda_j t)]}{2 \sinh^2(\lambda_j t)}.$$

The action between $\mathbf{x}_0$ and $\mathbf{x}$ in time t satisfies the equation $\frac{\partial}{\partial t} S = -E$ or

$$\frac{\partial}{\partial t} S = -\sum_{j=1}^{n} \frac{\lambda_j^2 [x_j^2 + (x_j^0)^2 - 2x_j x_j^0 \cosh(\lambda t)]}{2 \sinh^2(\lambda_j t)}$$

$$= \frac{\partial}{\partial t} \left[\sum_{j=1}^{n} \frac{\lambda_j}{2} (x_j^2 + (x_j^0)^2) \coth(\lambda_j t) - \sum_{j=1}^{n} \frac{\lambda_j x_j x_j^0}{\sinh(\lambda_j t)} \right].$$

Hence we shall choose

$$S = \sum_{j=1}^{n} \frac{\lambda_j}{2} \frac{1}{\sinh(\lambda_j t)} [(x_j^2 + (x_j^0)^2) \cosh(\lambda_j t) - 2x_j x_j^0]. \tag{10}$$

Let

$$S_j = \frac{\lambda_j}{2} \frac{1}{\sinh(\lambda_j t)} [(x_j^2 + (x_j^0)^2) \cosh(\lambda_j t) - 2x_j x_j^0]. \tag{11}$$

Then $S = S_1 + \cdots + S_n$ and $\partial_{x_j} S = \partial_{x_j} S_j$. Then Lemma 1 yields

$$\sum_{j=1}^{n} (\partial_{x_j} S)^2 = \sum_{j=1}^{n} (\partial_{x_j} S_j)^2 = \sum_{j=1}^{n} (\lambda_j^2 x_j^2 + 2E_j) = \sum_{j=1}^{n} \lambda_j^2 x_j^2 + 2E.$$

Hence

$$\sum_{j=1}^{n} \partial_{x_j}^2 S = \sum_{j=1}^{n} \partial_{x_j}^2 S_j = \sum_{j=1}^{n} \lambda_j \coth(\lambda_j t).$$

We shall look for a kernel of the form

$$K(x_0, x, t) = V(t) e^{kS(x_0, x, t)}, \quad k \in \mathbb{R}. \tag{12}$$

A computation similar to the one-dimensional case yields

$$\frac{\partial}{\partial t} K = e^{kS}(V'(t) - kEV(t)),$$

and

$$\partial_{x_j}^2 e^{kS} = e^{kS} k[k(\partial_{x_j} S)^2 + \partial_{x_j}^2 S],$$

and hence

$$\Delta_n e^{kS} = k e^{kS} \left\{ k \sum_{j=1}^n [\lambda_j^2 x_j^2 + 2E_j + \lambda_j \coth(\lambda_j t)] \right\}.$$

In order to find the kernel for the heat operator, we employ the multiplier method again. We shall consider the parabolic operator

$$P_n = \partial_t - \sum_{j=1}^n (\partial_{x_j}^2 - \alpha_j \lambda_j^2 x_j^2),$$

where α is a multiplier subject to be found later. Then

$$P_n K = e^{kS}[V'(t) - kEV(t)]$$

$$- k e^{kS} \left[k \left(\sum_{j=1}^n \lambda_j^2 x_j^2 + 2E_j \right) + \sum_{j=1}^n \lambda_j \coth(\lambda_j t) \right] V(t)$$

$$+ \sum_{j=1}^n \alpha_j \lambda_j^2 x_j^2 V(t) e^{kS}$$

$$= e^{kS} V(t) \left[\frac{V'(t)}{V(t)} - kE(1 + 2k) + \sum_{j=1}^n (\alpha_j - k^2) \lambda_j^2 x_j^2 - k \sum_{j=1}^n \lambda_j \coth(\lambda_j t) \right]$$

$$= e^{kS} V(t) \left[\frac{V'(t)}{V(t)} + \frac{na}{2} \coth(at) \right],$$

where we choose $k = -\frac{1}{2}$ and $\alpha_1 = \cdots = \alpha_n = \frac{1}{4}$. Let $\beta_j = \frac{\lambda_j}{2} \geq 0$ and choose $V(t)$ satisfying

$$\frac{V'(t)}{V(t)} = -\sum_{j=1}^n \lambda_j \coth(2\beta_j t), \quad t > 0.$$

Integrating yields $V(t) = \prod_{j=1}^n \frac{C_j}{\sinh^{1/2}(2\beta_j t)}$. Hence the fundamental solution for the operator $P_n = \partial_t - \sum_{j=1}^n (\partial_{x_j}^2 - \lambda_j^2 x_j^2)$ expressed in the form (12) is

$$K(x_0, x, t) = \prod_{j=1}^n \frac{C_j}{\sinh^{1/2}(2\beta_j t)} e^{-\sum_{j=1}^n \frac{2\beta_j}{4} \frac{1}{\sinh(2\beta_j t)} [(x_j^2 + (x_j^0)^2) \cosh(2\beta_j t) - 2x_j x_j^0]}$$

$$= \prod_{j=1}^{n} \frac{C_j}{\sinh^{1/2}(2\beta_j t)} e^{-\frac{1}{4t}\sum_{j=1}^{n} \frac{2\beta_j t}{\sinh(2\beta_j t)}[(x_j^2+(x_j^0)^2)\cosh(2\beta_j t)-2x_j x_j^0]}.$$

When $\beta_j \to 0$, $j = 1, \ldots, n$, we should obtain the kernel of the heat operator $\partial_t - \sum_{j=1}^{n} \partial_{x_j}^2$, which is

$$\frac{1}{(4\pi t)^{n/2}} e^{-\frac{1}{4t}|x-x_0|^2}, \quad t > 0.$$

By comparison, we obtain the value

$$C_j = \sqrt{\frac{\beta_j}{2\pi}}.$$

Theorem 3. Let $\beta_j \geq 0$ for $j = 1, \ldots, n$. The fundamental solution for the operator $P_n = \partial_t - \sum_{j=1}^{n}(\partial_{x_j}^2 - \beta_j^2 x_j^2)$ is

$$K(x_0, x, t) = \frac{1}{(4\pi t)^{n/2}} \left(\prod_{j=1}^{n} \frac{2\beta_j t}{\sinh(2\beta_j t)} \right)^{1/2}$$

$$\times \exp\left\{ -\frac{1}{4t} \sum_{j=1}^{n} \frac{2\beta_j t}{\sinh(2\beta_j t)} [(x_j^2 + (x_j^0)^2)\cosh(2\beta_j t) - 2x_j x_j^0] \right\}$$

$$= \frac{1}{(2\pi)^{\frac{n}{2}}} \left(\prod_{j=1}^{n} \frac{\beta_j}{\sinh(2\beta_j t)} \right)^{1/2}$$

$$\times \exp\left\{ -\sum_{j=1}^{n} \left[\frac{\beta_j(x_j - x_j^0)^2}{2\sinh(2\beta_j t)} + \frac{\beta_j(x_j^2 + (x_j^0)^2)}{2} \tanh(\beta_j t) \right] \right\}$$

for $t > 0$.

Theorem 3 recovers a result in [CGr] (see also [H2, Chapter 6]). In a similar way as in the one-dimensional case, choosing $\beta_j = -i\gamma_j$ yields the following result.

Theorem 4. Let $\gamma_j \geq 0$ for $j = 1, \ldots, n$. Then the fundamental solution for the operator $P = \partial_t - \sum_{j=1}^{n}(\partial_{x_j}^2 + \gamma_j^2 x_j^2)$ is

$$K(x_0, x, t) = \frac{1}{(4\pi t)^{n/2}} \left(\prod_{j=1}^{n} \frac{2\gamma_j t}{\sin(2\gamma_j t)} \right)^{1/2}$$

$$\times \exp\left\{ -\frac{1}{4t} \sum_{j=1}^{n} \frac{2\gamma_j t}{\sin(2\gamma_j t)} [(x_j^2 + (x_j^0)^2)\cos(2\gamma_j t) - 2x_j x_j^0] \right\}$$

$$= \frac{1}{(2\pi)^{\frac{n}{2}}} \left(\prod_{j=1}^{n} \frac{\gamma_j}{\sin(2\gamma_j t)} \right)^{1/2} e^{-\sum_{j=1}^{n} \left[\frac{\gamma_j (x_j - x_j^0)^2}{2\sin(2\gamma_j t)} + \frac{\gamma_j (x_j^2 + (x_j^0)^2)}{2} \tan(\gamma_j t) \right]}$$

for t > 0.

Remarks.

(1) We should point out here that the method we used in this section is based on Hamilton–Jacobi theory which provides deep insight into the geometry induced by the operator. Using this method, we not only can construct the heat kernel for the Hermite operator. We can also construct fundamental solutions for more general operators. For example, we can find the fundamental solution for the heat equation with quartic potential, i.e.,

$$P = \partial_t - \partial_x^2 - \frac{1}{4}\lambda^4 x^4 \quad \text{with } \lambda \geq 0.$$

This is a quite different behavior than the quadratic potential case (the Hermite operator), where there is only one energy and one solution between two given points. However, there are infinitely many energies associated to the quartic case. This makes the operator P very difficult to invert.

However, given any two points x_0 and x and a time $t > 0$, there is a sequence of energies $E_n = E_n(x_0, x, t)$ which have explicit representations. For each energy we associate an action $S_n = S_n(x_0, x, t)$, which satisfies the Hamilton–Jacobi equation

$$\partial_t S_n = -E_n(x_0, x, t).$$

We can show that

$$S_n \sim \frac{2}{3} \left(\frac{nK}{2a} \right)^4 \frac{1}{t^3} \quad \text{as } n \to \infty.$$

For each action S_n we associate a volume function V_n. Finally, we can write down the asymptotic expansion of the kernel

$$K(x_0, x, t) = \sum_{n=1}^{\infty} C_n V_n(t, x) e^{-\frac{1}{2} S_n}.$$

The constants C_n should be chosen such that

$$\sum_{n=1}^{\infty} C_n \lim_{t \searrow 0} V_n(t) \int_{\mathbb{R}} e^{-\frac{1}{2} S_n} \varphi(x)dx = \varphi(x_0)$$

for any compact supported function φ. Using this method, we may handle even more general potentials. Readers may consult [CC, Chapter 10].

(2) This method can be used to solve a more general operator such as

$$P = \partial_t - \partial_x^2 + \sqrt{2x}\partial_x.$$

The fundamental solution for the operator P is given by

$$K(x, x_0, t) = \frac{1}{\sqrt{2\pi t}} e^{-\frac{1}{2}\left(\frac{(x-x_0)^2}{2t} + \frac{x+x_0}{2}t - \frac{t^3}{24}\right)}. \tag{13}$$

This equation is derived from the famous Euler equation of compressible fluids (see, e.g., [L] and [LZ]).

(3) If one replaces t by it, the heat operator becomes a Schrödinger operator. The propagator for the Schrödinger operator $P = ih\partial_t + \frac{1}{2}h^2\partial_x - \frac{1}{2}\beta^2 x^2$ is

$$K(x, x_0, t, t_0) = \sqrt{\frac{ih\beta}{4\pi \sin(\alpha(t - t_0))}} e^{\frac{i}{2h}\left[\beta(x^2+x_0^2)\cot(\beta(t-t_0)) - \frac{2xx_0}{\sin(\beta(t-t_0))}\right]},$$

where $t - t_0 > 0$.

3 Hermite operator on the Heisenberg group

In this section, we will calculate the fundamental solution K_α for the Hermite operator on the Heisenberg group. First, note that for $j = 1, \ldots, n$,

$$\mathbf{X}_j^2 = \frac{\partial^2}{\partial x_j^2} + 4a_j^2 x_{j+n}^2 \frac{\partial^2}{\partial t^2} - 2a_j x_{j+n} \frac{\partial^2}{\partial x_j \partial t},$$

$$\mathbf{X}_{j+n}^2 = \frac{\partial^2}{\partial x_{j+n}^2} + 4a_j^2 x_j^2 \frac{\partial^2}{\partial t^2} + 2a_j x_j \frac{\partial^2}{\partial x_{j+n} \partial t}.$$

It follows that

$$\mathcal{H}_\alpha = \alpha + \sum_{j=1}^{2n}\left(\beta_j^2 x_j^2 - \frac{\partial^2}{\partial x_j^2}\right) - 4\sum_{j=1}^{2n} a_j^2 x_j^2 \frac{\partial^2}{\partial t^2}$$

$$+ 2\sum_{j=1}^{n} a_j \left(x_j \frac{\partial}{\partial x_{j+n}} - x_{j+n}\frac{\partial}{\partial x_j}\right)\frac{\partial}{\partial t} - i\lambda\frac{\partial}{\partial t}.$$

Here $a_j = a_{j+n}$ for $j = 1, \ldots, n$.

Since the operator $\mathbf{X}_j^2 + \mathbf{X}_{j+n}^2$ is invariant under rotations in the (x_j, x_{j+n}) plane, so the fundamental solution K_α is also invariant under rotations in the (x_j, x_{j+n}) plane (see [CT2] and [H2]). Therefore, we may assume that $K_\alpha(\mathbf{x}, t; \mathbf{y}, s)$ is independent of the rotational variables. It follows that

$$x_j \frac{\partial K_\alpha}{\partial x_{j+n}} - x_{j+n}\frac{\partial K_\alpha}{\partial x_j} = 0, \quad j = 1, \ldots, n.$$

Denote

$$\mathcal{R}_\alpha = \alpha + \sum_{j=1}^{2n} \beta_j^2 x_j^2 - \sum_{j=1}^{2n} \left[\frac{\partial^2}{\partial x_j^2} + 4a_j^2 x_j^2 \frac{\partial^2}{\partial t^2} \right] - i\lambda \frac{\partial}{\partial t}.$$

Taking the partial Fourier transform of $\mathcal{H}_\alpha$ with respect to the t-variable, one has

$$\widetilde{\mathcal{R}}_\alpha = \sum_{j=1}^{2n} \beta_j^2 x_j^2 - \sum_{j=1}^{2n} \left[\frac{\partial^2}{\partial x_j^2} - 4a_j^2 x_j^2 \tau^2 \right] + \alpha + \lambda\tau$$

$$= \alpha + \lambda\tau + \sum_{j=1}^{2n} \left[(4a_j^2 \tau^2 + \beta_j^2) x_j^2 - \frac{\partial^2}{\partial x_j^2} \right]$$

$$= \tilde{\alpha} + \sum_{j=1}^{2n} \left[\lambda_j^2 x_j^2 - \frac{\partial^2}{\partial x_j^2} \right].$$

This is exactly the Hermite operator on $\mathbb{R}^{2n}$ with parameter $\lambda_j = \sqrt{4a_j^2\tau^2 + \beta_j^2}$ and $\tilde{\alpha} = \alpha + \lambda\tau$. Recall that $a_j = a_{j+n}$ for $j = 1, \ldots, n$. Hence the exceptional set of $\mathcal{R}_\alpha$ is

$$\Lambda = \left\{ (\alpha, \lambda) : \alpha + \lambda\tau = -\sum_{j=1}^{2n} \sqrt{4a_j^2\tau^2 + \beta_j^2}(2k_j + 1); \ \mathbf{k} \in (\mathbb{Z}_+)^{2n} \right\}.$$

Now using Theorem 3 (see also [CC, Chapter 4] and [CT3, Theorem 2.1]), one can derive the fundamental solution of the Hermite operator by integration of the heat kernel.

Theorem 5. *For $\alpha \notin \{-\sum_{j=1}^{n} \lambda_j(2k_j+1), \mathbf{k} = (k_1, \ldots, k_n) \in (\mathbb{Z}_+)^n\}$, the Hermite operator $H_\alpha = \alpha - \Delta + \sum_{j=1}^{n} \lambda_j^2 x_j^2$ has fundamental solution*

$$K_\alpha(\mathbf{x}, \mathbf{y}) = \int_0^\infty e^{-\alpha s} P_s(\mathbf{x}, \mathbf{y}) ds,$$

where $P_s(\mathbf{x}, \mathbf{y})$ is defined as in Theorem 3.

Using Theorem 5, we know that for $(\alpha, \lambda) \notin \Lambda$, the fundamental solution for $\widetilde{\mathcal{R}}_\alpha$ is

$$\widetilde{K}_\alpha(\mathbf{x}, \tau; \mathbf{y}) = \frac{1}{(2\pi)^n} \int_0^\infty e^{-(\alpha+\lambda\tau)s} \left[\prod_{j=1}^{2n} \frac{\lambda_j}{\sinh(2\lambda_j s)} \right]^{\frac{1}{2}}$$

$$\times \exp\left\{ -\sum_{j=1}^{2n} \frac{\lambda_j}{2} \left[\frac{(x_j - y_j)^2}{\sinh(2\lambda_j s)} + (x_j^2 + y_j^2)\tanh(\lambda_j s) \right] \right\} ds.$$

Next, we can find the fundamental solution for the operator $\mathcal{H}_\alpha$ with singular point $(\mathbf{y}, 0)$ by taking the inverse Fourier transform with respect to τ. We note that $\lambda_j = \sqrt{4a_j^2\tau^2 + \beta_j^2}$ and it depends on τ:

$$K_\alpha(\mathbf{x}, t; \mathbf{y}, 0) = \frac{1}{2\pi} \int_{-\infty}^{\infty} e^{i\tau t} \widetilde{K}_\alpha(\mathbf{x}, \tau) d\tau$$

$$= \frac{1}{(2\pi)^{n+1}} \int_{-\infty}^{\infty} e^{i t \tau} \int_0^{\infty} e^{-(\alpha+\lambda\tau)s} \left[\prod_{j=1}^{2n} \frac{\lambda_j}{\sinh(2\lambda_j s)} \right]^{\frac{1}{2}}$$

$$\times \exp\left\{ -\sum_{j=1}^{2n} \frac{\lambda_j}{2} \left[\frac{(x_j - y_j)^2}{\sinh(2\lambda_j s)} + (x_j^2 + y_j^2)\tanh(\lambda_j s) \right] \right\} ds\, d\tau$$

$$= \frac{1}{(2\pi)^{n+1}} \int_{-\infty}^{\infty} \int_0^{\infty} e^{i t \tau - (\alpha+\lambda\tau)s} \left[\prod_{j=1}^{2n} \frac{(4a_j^2\tau^2 + \beta_j^2)^{\frac{1}{2}}}{\sinh(2s(4a_j^2\tau^2 + \beta_j^2)^{\frac{1}{2}})} \right]^{\frac{1}{2}}$$

$$\times \exp\left\{ -\sum_{j=1}^{2n} \frac{(4a_j^2\tau^2 + \beta_j^2)^{\frac{1}{2}}}{2} \left[\frac{(x_j - y_j)^2}{\sinh(2s(4a_j^2\tau^2 + \beta_j^2)^{\frac{1}{2}})} \right. \right.$$
$$\left. \left. + (x_j^2 + y_j^2)\tanh(s(4a_j^2\tau^2 + \beta_j^2)^{\frac{1}{2}}) \right] \right\} ds\, d\tau.$$

The fundamental solution at any point $(\mathbf{y}, s)$ is

$$K_\alpha(\mathbf{x}, t; \mathbf{y}, s) = K_\alpha(\mathbf{x}, t - s; \mathbf{y}, 0).$$

The above formula is very complicated and cannot be simplified in the general case.

3.1 Fundamental solution for the isotropic Heisenberg sub-Laplacian

It is very difficult to integrate the above integrals for arbitrary β_j and a_j. It can be simplified in some special cases. For example, if $\beta_j = 0$ for all j and $\alpha = 0$. We can find the fundamental solution of the sub-Laplacian $\mathcal{L}_\lambda$ with singular point at $(\mathbf{0}, 0)$:

$$K_\lambda(\mathbf{x}, t) = \frac{(n-1)!}{8\pi^{n+1}} \int_{-\infty}^{\infty} \left(\prod_{j=1}^{n} \frac{a_j}{\sinh(a_j s)} \right) \frac{e^{-\frac{\lambda}{4}s}}{\left(\sum_{j=1}^{2n} a_j x_j^2 \coth(a_j s) - it \right)^n} ds$$

$$= \frac{(n-1)!}{8\pi^{n+1}} \int_{-\infty}^{\infty} e^{-\frac{\lambda}{4}s} \frac{v(\tau) d\tau}{[\gamma(\mathbf{x}, \tau) - it]^n}.$$

Here

$$\gamma(\mathbf{x}, s) = \sum_{j=1}^{n} a_j(x_j^2 + x_{j+n}^2)\coth(a_j s) \quad \text{and} \quad v(s) = \prod_{j=1}^{n} \frac{a_j}{\sinh(a_j s)}.$$

This is the fundamental solution with singularity at the origin. Consequently, the fundamental solution with a singularity at an arbitrary point can be obtained by the Heisenberg translation.

When $a_j = 1$ for all $j = 1, \ldots, n$, then we may obtain an exact form for the kernel. In this case, the kernel will be

$$K_\lambda(\mathbf{x}, t) = \frac{(n-1)!}{8\pi^{n+1}} \int_{-\infty}^{\infty} \frac{1}{[\sinh(s)]^n} \frac{e^{-\frac{\lambda}{4}s}}{\left(\sum_{j=1}^{2n} x_j^2 \coth(s) - it\right)^n} ds$$

$$= \frac{(n-1)!}{8\pi^{n+1}} \int_{-\infty}^{\infty} \frac{e^{-\frac{\lambda}{4}s}}{(|\mathbf{x}|^2 \cosh(s) - it \sinh(s))^n} ds,$$

where $|\mathbf{x}|^2 = \sum_{j=1}^{2n} x_j^2$. Denote

$$r = (|\mathbf{x}|^4 + t^2)^{\frac{1}{4}} \quad \text{and} \quad e^{-i\varphi} = r^{-2}(|\mathbf{x}|^2 - it)$$

with $\varphi \in (-\frac{\pi}{2}, \frac{\pi}{2})$. Using the identity

$$\cosh(s + i\varphi) = \cosh(s) \cos \varphi + i \sinh(s) \sin \varphi,$$

one has

$$K_\lambda(\mathbf{x}, t) = \frac{(n-1)!}{8\pi^{n+1}} \int_{-\infty}^{\infty} \frac{e^{-\frac{\lambda}{4}s}}{[r^2 \cosh(s + i\varphi)]^n} ds. \tag{14}$$

Changing the contour, the formula (14) becomes

$$K_\alpha(\mathbf{x}, t) = \frac{(n-1)! \, \Gamma(n)}{8\pi^{n+1}} \frac{1}{r^{2n}} e^{i\frac{\lambda}{4}\varphi} \int_{-\infty}^{\infty} \frac{e^{-\frac{\lambda}{4}s}}{[\cosh(s)]^n} ds.$$

The above integral can be evaluated as follows:

$$K_\lambda(\mathbf{x}, t) = \frac{1}{8\pi^{n+1}} r^{-2n} e^{i\frac{\lambda}{4}\varphi} \Gamma\left(\frac{n+\lambda}{2}\right) \Gamma\left(\frac{n-\lambda}{2}\right)$$

$$= C_{n,\lambda}(|\mathbf{x}|^2 + it)^{-\left(\frac{n}{2} - \frac{\lambda}{8}\right)}(|\mathbf{x}|^2 - it)^{-\left(\frac{n}{2} + \frac{\lambda}{8}\right)}.$$

From the above formula, we know that the kernel can be extended from $|\operatorname{Re}(\lambda)| < n$ to the region $\mathbb{C} \setminus \mathcal{E}_\alpha$, where

$$\mathcal{E}_\alpha = \{\pm(n + 2k) : k \in \mathbb{Z}_+\}.$$

This coincides with the result obtained by Folland and Stein [FSt] except for a constant factor, since our vector fields have different constants in front of $\frac{\partial}{\partial t}$. In particular, when $\lambda = 0$, one has

$$K_0(\mathbf{x}, t) = C_n(|\mathbf{x}|^2 + it)^{-\frac{n}{2}}(|\mathbf{x}|^2 - it)^{-\frac{n}{2}} = C_n(|\mathbf{x}|^4 + t^2)^{-\frac{n}{2}}.$$

In fact, the above result can be generalized to the nonisotropic case by using Laguerre calculus; see, e.g., [B] and [BGr].

3.2 Connection with the Grusin operator

Consider the operator

$$\mathcal{L} = \frac{1}{2}\left[\left(\frac{\partial}{\partial x_1} + 2x_2\frac{\partial}{\partial t}\right)^2 + \left(\frac{\partial}{\partial x_2} - 2x_1\frac{\partial}{\partial t}\right)^2\right]$$

on the Heisenberg group. The fundamental solution for $\mathcal{L}$ with singularity at the origin is

$$K(x_1, x_2, t; 0, 0, 0) = \frac{1}{\pi^2}\int_{-\infty}^{\infty}\frac{\operatorname{csch}(2\tau)d\tau}{(x_1^2 + x_2^2)\coth(2\tau) - it}.$$

Set

$$x_1 = x, \qquad x_2 = z, \qquad y = 2x_1x_2 - 4t.$$

Then the operator $\mathcal{L}$ transforms to

$$\Delta = \frac{1}{2}\left[\left(\frac{\partial}{\partial x}\right)^2 + \left(x\frac{\partial}{\partial y} + \frac{\partial}{\partial z}\right)^2\right].$$

The operator Δ is translation invariant in y and z. Hence it suffices to have the singularity at $(x_0, 0, 0)$.

$$(x_0, 0, 0)^{-1} \cdot (x, z, 2x_1x_2 - 4y) = (x - x_0, z, 2(x_0 + x)z - 4y).$$

Therefore,

$$K(x, z, y; x_0, 0, 0) = \int_{\mathbb{R}}\frac{\pi^{-2}\operatorname{csch}(2\tau)d\tau}{[(x - x_0)^2 + z^2]\coth(2\tau) - i[2(x_0 + x)z - 4y]}.$$

Note that a parametrix for Δ_G may be obtained from a parametrix for Δ, using the Hadamard method of descent, by integrating the parametrix for Δ with respect to z. Recall

$$\int_{\mathbb{R}}\frac{d\lambda}{a\lambda^2 + b\lambda + c} = \frac{2\pi\operatorname{sgn}(a)}{\sqrt{4ac - b^2}}, \qquad \lambda \in \mathbb{R},$$

if $a \neq 0$ and $a\lambda^2 + b\lambda + c \neq 0$. Hence

$$\begin{aligned}
K(\mathbf{x}, \mathbf{x_0}, y) &= \int_{\mathbb{R}}\frac{2[\pi^2\sinh(2\tau)\cosh(2\tau)]^{-1/2}d\tau}{\sqrt{(\mathbf{x} - \mathbf{x_0})^2\coth(2\tau) + 4iy + (\mathbf{x} + \mathbf{x_0})^2\tanh(2\tau)}} \\
&= \frac{2}{\pi}\int_{\mathbb{R}}\frac{[\sinh(\tau)\cosh(\tau)]^{-1/2}d\tau}{\sqrt{(\mathbf{x} - \mathbf{x_0})^2\coth(\tau) + 4iy + (\mathbf{x} + \mathbf{x_0})^2\tanh(\tau)}}. \quad (15)
\end{aligned}$$

This is the fundamental solution for the step 2 Grusin operator

$$\Delta_G = \frac{1}{2}(X_1^2 + X_2^2),$$

where $X_1 = \partial_x$ and $X_2 = x\partial_y$. For further discussion, see [CCGrK1] and [CCGrK].

Acknowledgments Part of this article is based on the lecture presented by the second author at the conference on "A Celebration of Carlos Berenstein's Mathematics: Harmonic Analysis, Signal Processing and Complexity," which was held on May 17–29, 2004 at George Mason University, Fairfax, VA. We would like to thank Professor Daniele Struppa for the invitation. We would also like to thank the organizing committee for the warm hospitality during the conference.

References

[B]   R. Beals, A note on fundamental solutions, *Comm. Partial Differential Equations*, **24**-1–2 (1999), 369–376.

[BGGr]   R. Beals, B. Gaveau, and P. C. Greiner, Complex Hamiltonian mechanics and parametrics for subelliptic Laplacians I–III, *Bull. Sci. Math.*, **121** (1997), 1–36, 97–149, 195–259.

[BGr]   R. Beals and P. C. Greiner, *Calculus on Heisenberg Manifolds*, Annals of Mathematical Studies 119, Princeton University Press, Princeton, NJ, 1988.

[BCT]   C. Berenstein, D. C. Chang, and T. Tie, *Laguerre Calculus and Its Applications on the Heisenberg Group*, AMS/IP Series in Advanced Mathematics 22, International Press, Cambridge, MA, 2001.

[CC]   O. Calin, and D. C. Chang, *Geometric Mechanics on Riemannian Manifolds and Applications to PDEs*, Birkhäuser, Boston, 2004.

[CCGr]   O. Calin, D. C. Chang, and P. Greiner, On a step $2(k+1)$ subRiemannian manifold, *J. Geom. Anal.*, **15** (2004), 1–18.

[CCGrK1]   O. Calin, D. C. Chang, P. Greiner, and Y. Kannai, On the geometry induced by a Grusin operator, in L. Karp and L. Zalcman, eds., *Proceedings of the International Conference on Complex Analysis and Dynamical Systems* II, 2004, to appear.

[CCGrK]   O. Calin, D. C. Chang, P. Greiner, and Y. Kannai, Heat kernels for highly degenerate Grusin operators, in preparation, 2004.

[CGr]   D. C. Chang and P. Greiner, Analysis and geometry on Heisenberg groups, in C. S. Lin and S. T. Yau, eds., *Proceedings of Second International Congress of Chinese Mathematicians*, AMS/IP Series in Advanced Mathematics, International Press, Cambridge, MA, 2002.

[CT1]   D. C. Chang and J. Tie, Estimates for spectral projection operators of the sub-Laplacian on the Heisenberg group, *J. Anal. Math.*, **71** (1997), 315–347.

[CT2]   D. C. Chang and J. Tie, Estimates for powers of sub-Laplacian on the non-isotropic Heisenberg group, *J. Geom. Anal.*, **10** (2000), 653–678.

[CT3]   D. C. Chang and J. Tie, A note on Hermite and subelliptic operators, *Acta Math. Sinica*, 2004, to appear.

[FSt]   G. B. Folland and E. M. Stein, Estimates for the $\bar{\partial}_b$ complex and analysis on the Heisenberg group, *Comm. Pure Appl. Math.*, **27** (1974), 429–522.

[H1]   L. Hörmander, Hypoelliptic second-order differential equations, *Acta Math.*, **119** (1967), 147–171.

[H2]   L. Hörmander, *Riemannian Geometry*, Department of Mathematics, University of Lund, Lund, Sweden, 1990.

[L]   T. P. Liu, *Hyperbolic and Viscous Conservation Laws*, CBMS-NSF Regional Conference Series in Applied Mathematics 72, SIAM, Philadelphia, 2000.

[LZ]   Y. Zeng and T. P. Liu, *Large Time Behavior of Solutions for General Quasilinear Hyperbolic-Parabolic Systems of Conservation Laws*, Memoirs of the American Mathematical Society 125, American Mathematical Society, Providence, RI, 1997.

A Div-Curl Lemma in BMO on a Domain*

Der-Chen Chang[1], Galia Dafni[2], and Cora Sadosky[3]

[1] Department of Mathematics
Georgetown University
Washington, DC 20057
USA
chang@math.georgetown.edu
[2] Department of Mathematics and Statistics
Concordia University
Montreal, QC H3G-1M8
Canada
gdafni@mathstat.concordia.ca
[3] Department of Mathematics
Howard University
Washington, DC 20059
USA
csadosky@fac.howard.edu

Dedicated to Professor Carlos Berenstein on his 60th birthday.

Summary. Let $\Omega \subset \mathbf{R}^n$ be a Lipschitz domain. There are two BMO spaces, $\mathrm{BMO}_r(\Omega)$ and $\mathrm{BMO}_z(\Omega)$, which can be defined on Ω. The first part of this paper is a survey of some results for functions in these two spaces. The second part contains a div-curl-type lemma for $\mathrm{BMO}_r(\Omega)$ and $\mathrm{BMO}_z(\Omega)$.

Let f be a locally integrable function defined on $\mathbf{R}^n$, $n \geq 1$. Then f is said to have *bounded mean oscillation* if

$$\|f\|_{\mathrm{BMO}} := \sup \frac{1}{|Q|} \int_Q |f(x) - f_Q| dx < \infty,$$

where the supremum is taken over all cubes Q with sides parallel to the axes, $|Q|$ denotes the volume of the cube Q, and

---

* The research of the first author was partially supported by a William Fulbright Research Grant and U. S. Department of Defense Research Grant DAAH-0496-10301. The research of the second author was partially supported by the Natural Sciences and Engineering Research Council of Canada. The research of the third author was partially supported by U. S. Department of Energy grant DE-FG02-ER25341.

$$f_Q = \frac{1}{|Q|} \int_Q f(x)dx.$$

Modulo constants, $\| \cdot \|_{\mathrm{BMO}}$ defines a norm. The resulting Banach space, $\mathrm{BMO}(\mathbf{R}^n)$, was first introduced by F. John and L. Nirenberg [JN] in 1961, and has played an important role in the study of partial differential equations, in particular for endpoint results in Sobolev embeddings. It became the focus of attention in harmonic analysis when C. Fefferman [F] proved that BMO is the dual of the real Hardy space H^1 in 1971, and has since been studied extensively. An important example of the power of BMO is its use in the David–Journé $T(1)$ theorem for Calderón–Zygmund operators [DJ]. In studying of the boundedness of many types of operators, BMO often serves as a replacement for L^∞.

In recent years, the space BMO has also played a role in related fields, such as quasiconformal mappings and probability theory. For more background on BMO and Hardy spaces, see [ChSa], [Sa], and [S].

The first section of the paper reviews some notions of BMO spaces on domains in $\mathbf{R}^n$, and the corresponding Hardy spaces. The second section gives a version of the div-curl lemma of [CLMS] which is appropriate to these various BMO spaces.

1 The space BMO on a domain

Let Ω be an open connected subset of $\mathbf{R}^n$. By imposing the bounded mean oscillation condition only on cubes contained in Ω, Jones [Jo] defined a BMO space on Ω, which shall be denoted here by $\mathrm{BMO}_r(\Omega)$.

Definition 1. *A function g on Ω is in $\mathrm{BMO}_r(\Omega)$ if g is locally integrable and*

$$\sup_{Q \subset \Omega} \frac{1}{|Q|} \int_Q |g(x) - g_Q| dx \approx \|g\|_{\mathrm{BMO}_r(\Omega)} < \infty,$$

the supremum being taken over all cubes $Q \subset \Omega$ with sides parallel to the axes.

Jones [Jo] showed that, under a certain condition on the boundary $\partial\Omega$, which is satisfied in particular when the domain is Lipschitz, a function g is in $\mathrm{BMO}_r(\Omega)$ if and only if it admits an extension to a function G in $\mathrm{BMO}(\mathbf{R}^n)$, namely,

$$G|_\Omega = g \quad \text{and} \quad \|G\|_{\mathrm{BMO}(\mathbf{R}^n)} \le C_\Omega \|g\|_{\mathrm{BMO}_r(\Omega)}. \tag{1}$$

Here C_Ω is a constant depending only on Ω. Thus for a Lipschitz domain Ω one may consider $\mathrm{BMO}_r(\Omega)$ to be the space of restrictions to Ω of functions in $\mathrm{BMO}(\mathbf{R}^n)$.

One may ask whether it is possible to characterize those elements of $\mathrm{BMO}_r(\Omega)$ which can be extended by zero to functions in $\mathrm{BMO}(\mathbf{R}^n)$, or equivalently the subspace of $\mathrm{BMO}(\mathbf{R}^n)$ consisting of functions supported on $\overline{\Omega}$. Miyachi [Mi] achieved such a characterization, again under a geometric condition on the boundary, by imposing the bounded mean oscillation requirement on all cubes lying in the interior of Ω, as well as a stronger, bounded mean condition on cubes lying near the boundary. Following the terminology in [ChKS] (but using the constants in [Mi]), one has the following definitions.

Definition 2. *Let Ω be an open subset of $\mathbf{R}^n$. A cube $Q \subset \Omega$ (with sides parallel to the axes) is said to be of type (a) if $2Q \subset \Omega$. A cube is said to be of type (b) if $2Q \subset \Omega$ and $5Q \cap \Omega^c \neq \emptyset$.*

Here αQ indicates the cube with the same center as Q but α times the sidelength.

Definition 3. *Let Ω be an open subset of $\mathbf{R}^n$. A function g on Ω is in $\mathrm{BMO}_z(\Omega)$ if g is locally integrable and*

$$\sup_{type\ (a)\ cubes} \frac{1}{|Q|} \int_Q |g(x) - g_Q| dx + \sup_{type\ (b)\ cubes} \frac{1}{|Q|} \int_Q |g(x)| dx < \infty. \quad (2)$$

One may define $\|g\|_{\mathrm{BMO}_z(\Omega)}$ as the quantity on the left-hand side of (2).

For Ω a Lipschitz domain, it was shown by Miyachi [Mi] that $\mathrm{BMO}_z(\Omega)$ can be identified with the subspace $\{g \in \mathrm{BMO}(\mathbf{R}^n) : \mathrm{supp}(g) \subset \bar{\Omega}\}$, with equivalent norms.

Not surprisingly, in view of Fefferman's duality theorem, these two BMO spaces correspond to two kinds of real Hardy spaces on Ω. As in the papers [ChKS] and [ChDS], these two types of Hardy spaces will be called $H_r^p(\Omega)$ and $H_z^p(\bar{\Omega})$, $0 < p \leq 1$. The spaces $H_r^p(\Omega)$ were originally defined by Miyachi [Mi] as spaces of distributions on an arbitrary open set Ω in $\mathbf{R}^n$:

$$H_r^p(\Omega) = \{f \in (C_0^\infty)'(\Omega) : \mathrm{M}_{\phi,\Omega}(f) \in L^p(\Omega)\}, \quad (3)$$

where ϕ is a fixed C^∞ function with support in the unit ball $B(0, 1)$, $\int \phi = 1$, $\phi_t(y) = t^{-n}\phi(t^{-1}y)$, and

$$\mathrm{M}_{\phi,\Omega}(f)(x) = \sup_{0 < t < \mathrm{dist}(x, \partial\Omega)} |f * \phi_t(x)|.$$

Here the convolution $f * \phi_t(x)$ denotes the pairing $\langle f, \phi_t(x - \cdot)\rangle$. It was shown in [Mi] (under those geometric conditions on the boundary $\partial\Omega$ referred to above, which are satisfied in the case of a Lipschitz domain), that the space $H_r^p(\Omega)$ consists exactly of restrictions to Ω of distributions in the Hardy space $H^p(\mathbf{R}^n)$.

On the other hand, as discussed above for BMO, we may consider only those distributions in $H^p(\mathbf{R}^n)$ which vanish outside Ω, namely, whose support is contained in the closure $\bar{\Omega}$. In the case where $\bar{\Omega}$ is replaced by an arbitrary closed set satisfying a certain geometric condition, these Hardy spaces were defined by Jonsson, Sjögren, and Wallin [JSW]. For a Lipschitz domain Ω, this gives the following definition of the spaces $H_z^p(\bar{\Omega})$:

$$H_z^p(\bar{\Omega}) = \{f \in H^p(\mathbf{R}^n) : f = 0 \quad \text{on} \quad \mathbf{R}^n \setminus \bar{\Omega}\}.$$

When $p = 1$, the elements of H_r^1 and H_z^1 are functions and therefore cannot be supported on $\partial\Omega$. This justifies the simplified notation $H_z^1(\Omega)$, instead of $H_z^1(\bar{\Omega})$.

Using a constructive method, one may derive atomic decompositions for these two Hardy spaces. For the sake of simplicity, and since this is the only case of relevance to BMO, let $p = 1$. Again, one needs to distinguish between the behavior on cubes in the interior and on cubes near the boundary.

A function a is called an H^1 atom if it is supported in a cube Q, has norm $\|a\|_{L^2}$ bounded by $|Q|^{-1/2}$, and satisfies the cancellation condition

$$\int_Q a(x)dx = 0.$$

If a is supported in a type (a) cube in Ω, it will be called a type (a) atom. Call a a type (b) atom if it is supported in a type (b) cube and satisfies the size condition $\|a\|_{L^2} \le |Q|^{-1/2}$, but not necessarily the vanishing moment condition.

The following theorem is a combination of the atomic decompositions proved in [JSW] (part (b)), [Mi] (part (a)), and [ChKS] for the case $p = 1$:

Theorem 1.

(a) *A function f belongs to $H_r^1(\Omega)$ if and only if f has an atomic decomposition:*

$$f = \sum \lambda_j a_j + \sum \mu_k b_k,$$

with the a_j type (a) atoms, the b_k type (b) atoms, and the coefficients satisfying

$$\sum |\lambda_j| + \sum |\mu_j| < \infty.$$

(b) *A function f belongs to $H_z^1(\Omega)$ if and only if f has an atomic decomposition as follows:*

$$f = \sum \lambda_Q a_Q,$$

the sum being taken over a countable collection of cubes $Q \subset \Omega$, with $\{a_Q\}$ a sequence of H^1 atoms and $\{\lambda_Q\}$ satisfying

$$\sum |\lambda_Q| < \infty.$$

This gives the following duality results (see [JSW], [Mi], and [Ch]).

Theorem 2. *Let Ω be a Lipschitz domain in $\mathbf{R}^n$.*

(a) *If $g \in \mathrm{BMO}_r(\Omega)$, then there exists a unique linear functional $\mathcal{L}$ in the dual space of $H_z^1(\Omega)$ such that*

$$\mathcal{L}(f) = \int_\Omega f(x)g(x)dx \tag{4}$$

for all $f \in H_z^1(\Omega)$. Conversely, if $\mathcal{L}$ is in the dual space of $H_z^1(\Omega)$, then there exists a unique $g \in \mathrm{BMO}_r(\Omega)$ such that (4) holds. The correspondence $\mathcal{L} \leftrightarrow g$ given by (4) is a Banach space isomorphism between $\mathrm{BMO}_r(\Omega)$ and the dual space of $H_z^1(\Omega)$.

(b) If $g \in \mathrm{BMO}_z(\Omega)$, then there exists a unique linear functional $\mathcal{L}$ in the dual space of $H_r^1(\Omega)$ such that (4) holds for all $f \in H_r^1(\Omega)$. Conversely, if $\mathcal{L}$ is in the dual space of $H_r^1(\Omega)$, then there exists a unique $g \in \mathrm{BMO}_z(\Omega)$ such that (4) holds. The correspondence $\mathcal{L} \leftrightarrow g$ given by (4) is a Banach space isomorphism between $\mathrm{BMO}_z(\Omega)$ and the dual space of $H_r^1(\Omega)$.

It should be remarked that since the main purpose of the papers [ChKS] and [ChDS] was the study of H^p regularity properties of boundary value problems for an elliptic operator on a smooth domain in $\mathbf{R}^n$, the Hardy spaces used were the local Hardy spaces $h^p(\mathbf{R}^n)$ developed by Goldberg [Go]. These can be defined, for example, via an atomic decomposition in which atoms supported in large cubes ($|Q| > 1$) need not satisfy a cancellation condition. The definitions above can be suitably modified to obtain the spaces $h_r^p(\Omega)$, $h_z^p(\Omega)$, and the corresponding local BMO spaces $\mathrm{bmo}_r(\Omega)$ and $\mathrm{bmo}_z(\Omega)$ (see [Ch]).

2 A div-curl-type theorem

In [CLMS], Coifman, Lions, Meyer, and Semmes showed that for exponents p, q with $1 < p < \infty$, $\frac{1}{p} + \frac{1}{q} = 1$, and vector fields $\mathbf{V}$ in $L^p(\mathbf{R}^n, \mathbf{R}^n)$, $\mathbf{W}$ in $L^q(\mathbf{R}^n, \mathbf{R}^n)$ with $\mathrm{div}\,\mathbf{V} = 0$, $\mathrm{curl}\,\mathbf{W} = 0$ in the sense of distributions, the scalar (dot) product $\mathbf{V} \cdot \mathbf{W}$ belongs to $H^1(\mathbf{R}^n)$. Moreover,

$$\|\mathbf{V} \cdot \mathbf{W}\|_{H^1} \leq C \|\mathbf{V}\|_{L^p} \|\mathbf{W}\|_{L^q}. \tag{5}$$

In the same paper [CLMS], they also proved that for $g \in \mathrm{BMO}(\mathbf{R}^n)$,

$$\|g\|_{\mathrm{BMO}} \approx \sup_{\mathbf{V}, \mathbf{W}} \int_{\mathbf{R}^n} g\mathbf{V} \cdot \mathbf{W}, \tag{6}$$

where the supremum is taken over all vector fields $\mathbf{V}, \mathbf{W}$ in $L^2(\mathbf{R}^n, \mathbf{R}^n)$, $\|\mathbf{V}\|_{L^2}$, $\|\mathbf{W}\|_{L^2} \leq 1$, satisfying $\mathrm{div}\,\mathbf{V} = 0$, $\mathrm{curl}\,\mathbf{W} = 0$ in the sense of distributions on $\mathbf{R}^n$. Here and below, one must obviously consider only real-valued functions g in BMO.

Recently, several variants of these results have been proved for Hardy spaces on domains. For a version in $H_z^1(\Omega)$, see [HLMZ] and [ART]. For results in the context of differential forms, see [HLMZ], [LM], and [T, Section 3.8]. The latter also contains generalizations of the nonhomogeneous version of (5) given in [CLMS] with respect to h_{loc}^1, a localized version of the Hardy space H^1. Nonhomogeneous div-curl lemmas for the local Hardy space h^1 of Goldberg [Go], and for $h_r^1(\Omega)$, may be found in [D].

The goal of this section is to prove analogues of (6) for functions in $\mathrm{BMO}_r(\Omega)$ and $\mathrm{BMO}_z(\Omega)$. Before proceeding with this, one needs to understand the different possible meanings of the vanishing divergence and curl conditions when applied to vector fields on a domain Ω.

Let Ω be an open subset of $\mathbf{R}^n$, and suppose $\mathbf{v} = (v_1, \ldots, v_n) \in L^p(\Omega, \mathbf{R}^n)$. One says that $\mathrm{div}\,\mathbf{v} = 0$ in the sense of distributions on Ω if

$$\int_\Omega \mathbf{v} \cdot \nabla \varphi = 0 \tag{7}$$

for all $\varphi \in C_0^\infty(\Omega)$ (i.e., smooth functions with compact support in Ω), or that curl $\mathbf{v} = 0$ in the sense of distributions on Ω if

$$\int_\Omega v_j \frac{\partial \varphi}{\partial x_i} - v_i \frac{\partial \varphi}{\partial x_j} = 0 \tag{8}$$

for all $i, j \in \{1, \ldots, n\}$ and all $\varphi \in C_0^\infty(\Omega)$. If the components of $\mathbf{v}$ are sufficiently smooth, this is equivalent to the classical notion of vanishing divergence and curl via integration by parts.

One would like to also consider the conditions given by equations (7) and (8) in the case when the test functions do not have compact support in Ω. This is equivalent to saying that the vector field

$$\mathbf{V} = \begin{cases} \mathbf{v} & \text{in } \Omega, \\ \mathbf{0} & \text{in } \mathbf{R}^n \setminus \Omega \end{cases} \tag{9}$$

satisfies div $\mathbf{V} = 0$ (respectively, curl $\mathbf{V} = 0$) in the sense of distributions on $\mathbf{R}^n$.

If $\mathbf{v}$ has smooth components, and the boundary $\partial \Omega$ of Ω is sufficiently smooth, the integration by parts in (7), when φ does not have compact support in Ω, implicitly incorporates the boundary condition $\mathbf{n} \cdot \mathbf{v} = 0$ on $\partial \Omega$, where $\mathbf{n} = (\eta_1, \ldots, \eta_n)$ denotes the outward unit normal vector. Similarly, in (8), if φ is not of compact support in Ω, the implicit boundary conditions are

$$v_j \eta_i = v_i \eta_j,$$

meaning that $\mathbf{v}$ is colinear with $\mathbf{n}$ on $\partial \Omega$. It should be pointed out that in the latter case, since (8) shows that curl $\mathbf{V} = 0$ in $\mathbf{R}^n$ for the extension $\mathbf{V}$ of $\mathbf{v}$ defined in (9), and $\mathbf{R}^n$ is simply connected, one immediately obtains a potential π so that $\mathbf{V} = \nabla \pi$ and hence $\mathbf{v} = \nabla \pi$ in Ω. (In [ART], condition (11) below was replaced by the assumption that $\mathbf{v} = \nabla \pi$ weakly in Ω).

Thus for a general vector field $\mathbf{v} \in L^p(\Omega, \mathbf{R}^n)$, with Ω, say, a Lipschitz domain in $\mathbf{R}^n$, following the usual notation for the statement of the Neumann problem on Ω, write

$$\begin{cases} \text{div } \mathbf{v} = 0 & \text{in } \Omega, \\ \mathbf{n} \cdot \mathbf{v} = 0 & \text{on } \partial \Omega \end{cases} \tag{10}$$

to indicate that (7) holds for all $\varphi \in C_0^\infty(\mathbf{R}^n)$, and

$$\begin{cases} \text{curl } \mathbf{v} = 0 & \text{in } \Omega, \\ \mathbf{n} \parallel \mathbf{v} & \text{on } \partial \Omega \end{cases} \tag{11}$$

to indicate that (8) holds for all $i, j \in \{1, \ldots, n\}$ and all $\varphi \in C_0^\infty(\mathbf{R}^n)$.

In order to prove an analogue of (6) for $\mathrm{BMO}_z(\Omega)$, one needs the following version of the div-curl lemma for $H_r^1(\Omega)$. This can be done in any open subset of $\mathbf{R}^n$.

The proof is a simplified version of the argument used to prove a nonhomogeneous div-curl lemma for $h_r^1(\Omega)$ in [D], which in turn is adapted from the proof of [CLMS, Lemma II.1].

Theorem 3. *Suppose* $\mathbf{v}$ *and* $\mathbf{w}$ *are vector fields on an open set* $\Omega \subset \mathbf{R}^n$, *satisfying*

$$\mathbf{v} \in L^p(\Omega, \mathbf{R}^n), \qquad \mathbf{w} \in L^q(\Omega, \mathbf{R}^n), \qquad 1 < p < \infty, \qquad \frac{1}{p} + \frac{1}{q} = 1,$$

and

$$\operatorname{div} \mathbf{v} = 0, \qquad \operatorname{curl} \mathbf{w} = 0$$

in the sense of distributions on Ω. *Then* $\mathbf{v} \cdot \mathbf{w}$ *belongs to the Hardy space* $H_r^1(\Omega)$ *with*

$$\|\mathbf{v} \cdot \mathbf{w}\|_{H_r^1(\Omega)} \leq C \|\mathbf{v}\|_{L^p(\Omega)} \|\mathbf{w}\|_{L^q(\Omega)}. \tag{12}$$

Proof. Given a point $x \in \Omega$ and a radius $t < \operatorname{dist}(x, \partial\Omega)$ (so that the ball $B(x, t)$ lies inside Ω), one can find a function π in the Sobolev space $W^{1,q}(B(x, t))$ with $\nabla\pi = \mathbf{w}$. Without loss of generality, assume $\int_{B(x,t)} \pi = 0$.

Then for $\phi \in C_0^\infty(B(0, 1))$ with $\int \phi = 1$, setting $\phi_t(y) = t^{-n}\phi(t^{-1}y)$, one can use the fact that $\operatorname{supp}(\phi_t(x - \cdot)) \subset B(x, t) \subset \Omega$, and $\operatorname{div} \mathbf{v} = 0$ in the sense of distributions on Ω, to write

$$-\phi_t * (\mathbf{v} \cdot \mathbf{w})(x) = \int_{B(x,t)} t^{-(n+1)} \nabla\phi(t^{-1}(x - y)) \cdot \mathbf{v}(y)\pi(y)dy.$$

(This can be verified by approximating π with smooth functions.)

The Sobolev–Poincaré inequality in $B(x, t)$, together with the fact that $\int_{B(x,t)} \pi = 0$, gives (see the proof of [CLMS, Lemma II.1]):

$$|\phi_t * (\mathbf{v} \cdot \mathbf{w})(x)| \leq C \left(t^{-n} \int_{B_t^x} |\mathbf{v}|^\alpha \right)^{1/\alpha} \left(t^{-n} \int_{B_t^x} |\nabla\pi_t^x|^\beta \right)^{1/\beta}$$

$$\leq C(\mathrm{M}(|\mathbf{v}|^\alpha)(x))^{1/\alpha} (\mathrm{M}(|\mathbf{w}|^\beta)(x))^{1/\beta}$$

for some α, β with $1 < \alpha < p$, $1 < \beta < q$ and $1/\alpha + 1/\beta = 1 + 1/n$. Here M denotes the Hardy–Littlewood maximal function, applied to the functions $|\mathbf{v}|^\alpha$ and $|\mathbf{w}|^\beta$, which are extended to $\mathbf{R}^n$ by setting them equal to zero outside Ω. Note that the constant does not depend on t or x.

The boundedness of the maximal function on $L^r(\mathbf{R}^n)$, $r > 1$, results in the following estimate:

$$\int_\Omega \sup_{0 < t < \operatorname{dist}(x, \partial\Omega)} |\phi_t * (\mathbf{v} \cdot \mathbf{w})(x)| dx$$

$$\leq C \left(\int_\Omega (\mathrm{M}(|\mathbf{v}|^\alpha)(x))^{p/\alpha} dx \right)^{1/p} \left(\int_\Omega (\mathrm{M}(|\mathbf{w}|^\beta)(x))^{q/\beta} dx \right)^{1/q}$$

$$\leq C \|\mathbf{v}\|_{L^p(\Omega)} \|\mathbf{w}\|_{L^q(\Omega)},$$

which, by (3), shows $\mathbf{v} \cdot \mathbf{w} \in H_r^1(\Omega)$, and (12) holds. $\qquad\square$

Theorem 4. *Let $\Omega \subset \mathbf{R}^n$ be a Lipschitz domain. Then we have the following results.*

(a) *If $g \in \mathrm{BMO}_z(\Omega)$, then*

$$\|g\|_{\mathrm{BMO}_z} \approx \sup_{\mathbf{v},\mathbf{w}} \int_\Omega g\mathbf{v} \cdot \mathbf{w}, \tag{13}$$

where the supremum is taken over all vector fields $\mathbf{v}, \mathbf{w} \in L^2(\Omega, \mathbf{R}^n)$, $\|\mathbf{v}\|_{L^2(\Omega)} \le 1$, $\|\mathbf{w}\|_{L^2(\Omega)} \le 1$, satisfying $\mathrm{div}\,\mathbf{v} = 0$ and $\mathrm{curl}\,\mathbf{w} = 0$ in the sense of distributions on Ω.

(b) *If $g \in \mathrm{BMO}_r(\Omega)$, then*

$$\|g\|_{\mathrm{BMO}_r} \approx \sup_{\mathbf{v},\mathbf{w}} \int_\Omega g\mathbf{v} \cdot \mathbf{w},$$

the supremum being taken over all vector fields $\mathbf{v}$ and $\mathbf{w}$ in $L^2(\Omega, \mathbf{R}^n)$ with $\|\mathbf{v}\|_{L^2(\Omega)} \le 1$, $\|\mathbf{w}\|_{L^2(\Omega)} \le 1$ and satisfying conditions (10) and (11).

Proof.

(a) Let $g \in \mathrm{BMO}_z(\Omega)$ and take $\mathbf{v}, \mathbf{w} \in L^2(\Omega, \mathbf{R}^n)$, $\|\mathbf{v}\|_{L^2(\Omega)} \le 1$, $\|\mathbf{w}\|_{L^2(\Omega)} \le 1$, with $\mathrm{div}\,\mathbf{v} = 0$ and $\mathrm{curl}\,\mathbf{w} = 0$ in the sense of distributions on Ω. By Theorem 3, the dot product $\mathbf{v} \cdot \mathbf{w}$ belongs to $H_r^1(\Omega)$ with norm bounded by a constant. The duality of $\mathrm{BMO}_z(\Omega)$ with $H_r^1(\Omega)$ (Theorem 2(b)) then gives

$$\int_\Omega g\,\mathbf{v} \cdot \mathbf{w} \le C\|g\|_{\mathrm{BMO}_z}.$$

Conversely, by the remark following Definition 3, the extension

$$G = \begin{cases} g & \text{in } \Omega, \\ 0 & \text{in } \mathbf{R}^n \setminus \Omega \end{cases}$$

of g to $\mathbf{R}^n$ is in $\mathrm{BMO}(\mathbf{R}^n)$ with $\|G\|_{\mathrm{BMO}} \approx \|g\|_{\mathrm{BMO}_z}$. Hence, by (6), one has

$$\|g\|_{\mathrm{BMO}_z} \approx \sup_{\mathbf{V},\mathbf{W}} \int_{\mathbf{R}^n} g\mathbf{V} \cdot \mathbf{W} \approx \sup_{\mathbf{V},\mathbf{W}} \int_\Omega g\mathbf{V}|_\Omega \cdot \mathbf{W}|_\Omega,$$

where the supremum is taken over all vector fields $\mathbf{V}$, $\mathbf{V} \in L^2(\mathbf{R}^n, \mathbf{R}^n)$, $\|\mathbf{V}\|_{L^2} \le 1$, $\|\mathbf{W}\|_{L^2} \le 1$, satisfying $\mathrm{div}\,\mathbf{V} = 0$ and $\mathrm{curl}\,\mathbf{W} = 0$ in the sense of distributions on $\mathbf{R}^n$. For such vector fields, clearly the restrictions $\mathbf{v} = \mathbf{V}|_\Omega$, $\mathbf{w} = \mathbf{W}|_\Omega$ satisfy the same conditions in Ω, and therefore (13) holds.

(b) Since $g \in \mathrm{BMO}_r(\Omega)$, by (1), there exists $G \in \mathrm{BMO}(\mathbf{R}^n)$ such that $G|_\Omega = g$ and

$$\|G\|_{\mathrm{BMO}} \le C\|g\|_{\mathrm{BMO}_r}.$$

Given $\mathbf{v}$ and $\mathbf{w}$ in $L^2(\Omega, \mathbf{R}^n)$, define the zero extensions

$$V = \begin{cases} v & \text{in } \Omega, \\ 0 & \text{in } \mathbf{R}^n \setminus \Omega \end{cases}$$

and

$$W = \begin{cases} w & \text{in } \Omega, \\ 0 & \text{in } \mathbf{R}^n \setminus \Omega. \end{cases}$$

Conditions (10) and (11) are equivalent to the fact that div $V = 0$, curl $W = 0$ in the sense of distributions on $\mathbf{R}^n$. Therefore, by duality Theorem 2(a) and inequality (5), one has

$$
\begin{aligned}
\int_\Omega g v \cdot w = \int_{\mathbf{R}^n} G V \cdot W \\
\le \|G\|_{\mathrm{BMO}} \|V \cdot W\|_{H^1} \\
\le C \|g\|_{\mathrm{BMO}_r} \|V\|_{L^2} \|W\|_{L^2} \\
= C \|g\|_{\mathrm{BMO}_r} \|v\|_{L^2} \|w\|_{L^2} \\
\le C \|g\|_{\mathrm{BMO}_r}.
\end{aligned}
$$

This shows that

$$\sup_{v,w} \int_\Omega g v \cdot w \le C \|g\|_{\mathrm{BMO}_r}.$$

It remains to prove the other direction, i.e.,

$$\|g\|_{\mathrm{BMO}_r} \le C' \sup_{v,w} \int_\Omega g v \cdot w.$$

If Q is a type (a) cube, namely, $\widetilde{Q} = 2Q \subset \Omega$, then one can use the following estimate from the proof of [CLMS, Theorem III.2]:

$$\left(\frac{1}{|Q|} \int_Q |g(x) - g_Q|^2 dx \right)^{1/2} \le C_n \sup \int g v \cdot w, \tag{14}$$

where the supremum is taken over all vector fields $v, w \in C_0^\infty(\widetilde{Q})$ with $\|v\|_{L^2} \le 1$, $\|w\|_{L^2} \le 1$ and div $v = 0$, curl $w = 0$, since such vector fields automatically satisfy the boundary conditions $v \perp n$ on $\partial\Omega$, $w \parallel n$ on $\partial\Omega$.

Now note that in order to estimate the BMO$_r$ norm on a Lipschitz domain Ω, it suffices to bound the mean oscillation on type (a) cubes. In fact, this can be deduced from [Jo], where the extension of a BMO$_r(\Omega)$ function to a function in BMO$(\mathbf{R}^n)$ is proved using the Whitney decomposition of Ω, in which the cubes are of a size proportional to their distance from the boundary.

Another way to observe this is using the duality between BMO$_r(\Omega)$ and $H_z^1(\Omega)$ (Theorem 2(a)). The proof of the atomic decomposition of $H_z^1(\Omega)$ (Theorem 1(b)) given in [ChKS] can be slightly modified (by choosing a reproducing kernel with support in a smaller cone) so that the resulting atoms are supported in type (a) cubes, although in this case instead of $2Q \subset \Omega$ one only obtains $\alpha Q \subset \Omega$ for some constant $\alpha > 1$ depending on Ω. In order to show that integration against a function g gives a bounded linear functional on $H_z^1(\Omega)$, and hence g is in $BMO_r(\Omega)$, it then suffices to estimate the mean oscillation on these type (a) cubes. Clearly, the argument in [CLMS] resulting in the estimate (14) can be adapted to the case where 2 is replaced by α.

A third argument is just a simple geometric construction showing that the mean oscillation of g on any cube $Q \subset \Omega$ can be estimated by the supremum of its mean oscillation on subcubes Q' of Q whose distance from the boundary of Q is proportional to their size (i.e., Whitney cubes with respect to Q). Such cubes Q' are then automatically type (a) cubes with respect to Ω. □

Using [CLMS, Lemmas III.1 and III.2], and the duality Theorem 2, one obtains as a corollary the following decomposition.

Theorem 5. *Let* $\Omega \subset \mathbf{R}^n$ *be a Lipschitz domain.*

(a) *For a function* f *in* $H_r^1(\Omega)$, *there exist two sequences* $\{\mathbf{v}_k\}, \{\mathbf{w}_k\} \subset L^2(\Omega, \mathbf{R}^n)$ *with* $\|\mathbf{v}_k\|_{L^2}, \|\mathbf{w}_k\|_{L^2} \leq 1$, *satisfying* $\mathrm{div}\, \mathbf{v}_k = 0$ *and* $\mathrm{curl}\, \mathbf{w}_k = 0$ *in the sense of distributions on* Ω, *so that*

$$f = \sum_{k=1}^{\infty} \lambda_k \mathbf{v}_k \cdot \mathbf{w}_k.$$

Moreover, one has $\sum_{k=1}^{\infty} |\lambda_k| < \infty$.

(b) *For a function* $f \in H_z^1(\Omega)$, *there exist two sequences* $\{\mathbf{v}_k\}, \{\mathbf{w}_k\} \subset L^2(\Omega, \mathbf{R}^n)$ *with* $\|\mathbf{v}_k\|_{L^2}, \|\mathbf{w}_k\|_{L^2} \leq 1$, *satisfying conditions* (10) *and* (11), *respectively, so that*

$$f = \sum_{k=1}^{\infty} \lambda_k \mathbf{v}_k \cdot \mathbf{w}_k.$$

Moreover, one has $\sum_{k=1}^{\infty} |\lambda_k| < \infty$.

References

[ART]   P. Auscher, E. Russ, and P. Tchamitchian, Hardy-Sobolev spaces on strongly Lipschitz domains of $\mathbf{R}^n$, preprint, 2003.

[Ch]   D. C. Chang: The dual of Hardy spaces on a bounded domain in $\mathbf{R}^n$, *Forum Math.*, **6** (1994), 65–81.

[ChDS]   D. C. Chang, G. Dafni, and E. M. Stein, Hardy spaces, BMO, and boundary value problems for the Laplacian on a smooth domain in $\mathbf{R}^n$, *Trans. Amer. Math. Soc.*, **351** (1999), 1605–1661.

[ChKS]   D. C. Chang, S. G. Krantz, and E. M. Stein, H^p Theory on a smooth domain in $\mathbf{R}^N$ and elliptic boundary value problems, *J. Functional Anal.*, **114** (1993), 286–347.

[ChSa]   D. C. Chang and C. Sadosky, Functions of bounded mean oscillation, *Taiwanese J. Math.*, to appear, 2005.

[CLMS] R. Coifman, P.-L. Lions, Y. Meyer, and S. Semmes, Compensated compactness and Hardy spaces, *J. Math. Pures Appl.*, **72** (1993), 247–286.

[D]   G. Dafni, Nonhomogeneous div-curl lemmas and local Hardy spaces, *Adv. Differential Equations*, **10** (2005), 505–526.

[DJ]   G. David and J.-L. Journé, A boundedness criterion for generalized Calderón–Zygmund operators, *Ann. Math.*, **120** (1984), 371–397.

[F]   C. Fefferman, Characterizations of bounded mean oscillations, *Bull. Amer. Math. Soc.*, **77** (1971), 587–588.

[FS]   C. Fefferman and E. M. Stein, H^p spaces of several variables, *Acta Math.*, **129** (1972), 137–193.

[Go]   D. Goldberg, A local version of real Hardy spaces, *Duke Math. J.*, **46** (1979), 27–42.

[HLMZ] J. Hogan, C. Li, A. McIntosh, and K. Zhang, Global higher integrability of Jacobians on bounded domains, *Ann. Inst. H. Poincaré Anal. Non Linéaire*, **17** (2000), 193–217.

[Jo]   P. W. Jones, Extension theorems for BMO, *Indiana Univ. Math. J.*, **29** (1980), 41–66.

[JN]   F. John and L. Nirenberg, On functions of bounded mean oscillation, *Comm. Pure Appl. Math.*, **14** (1961), 415–426.

[JSW]   A. Jonsson, P. Sjögren, and H. Wallin, Hardy and Lipschitz spaces on subsets of $\mathbf{R}^n$, *Stud. Math.*, **80** (1984), 141–166.

[LM]   Z. Lou and A. McIntosh, Hardy spaces of exact forms on Lipschitz domains in $\mathbf{R}^N$, *Indiana Univ. Math. J.*, **53** (2004), 583–611.

[Mi]   A. Miyachi, H^p spaces over open subsets of $\mathbf{R}^n$, *Stud. Math.*, **95** (1990), 205–228.

[Sa]   C. Sadosky, *Interpolation of Operators and Singular Integrals*, Marcel Dekker, New York, Basel, 1979.

[S]   E. M. Stein, *Harmonic analysis: Real-Variable Methods, Orthogonality, and Oscillatory Integrals*, Princeton University Press, Princeton, NJ, 1993.

[T]   M. E. Taylor, *Tools for PDE: Pseudodifferential Operators, Paradifferential Operators, and Layer Potentials*, Mathematical Surveys and Monographs 81, American Mathematical Society, Providence, RI, 2000.

Subharmonic Functions on Discrete Structures

Christer O. Kiselman

Uppsala University
P. O. Box 480
SE-751 06 Uppsala
Sweden
kiselman@math.uu.se

Dedicated to Carlos Berenstein and Daniele Struppa.

Abstract. We define the Laplacian on an arbitrary set with a not necessarily symmetric weight function and discuss the Dirichlet problem and other classical topics in this setting.

1 Introduction

The problem of describing the shape of a three-dimensional object is important in many applications. Images in medicine and industry are often three-dimensional nowadays.

One should be able to store the description of a shape in a computer and be able to compare it with other shapes, using some measure of likeness. One approach to shape description is to introduce a triangulation of the surface of the object and then map this triangulation to a sphere. The position of a point on the surface is then a function on the sphere, and can be expanded in terms of spherical harmonics. This approach, initiated by Brechbühler et al. (1995), was the background of my talk at the Berenstein conference, and it leads to the study of harmonic, or more generally subharmonic, functions on a graph or rather a directed graph with weight functions that need not be symmetric; indeed the Dirichlet problem is not interesting in the symmetric case.

It turns out that the values of harmonic functions often cluster together in an undesirable way, and to get rid of this clustering is a special problem of importance in the shape-description project of Ola Weistrand (ms.). This means that, although there is a homeomorphism from the surface of the object to a sphere, the modulus of continuity of the inverse mapping is terribly large. There are various remedies, one being to use different weights in the definition of harmonicity.

This paper is an introduction to the study of harmonic and subharmonic functions on discrete structures. The Dirichlet problem will be studied and explicit solutions in some simple cases will be given. In other cases, however, explicit formulas corresponding to well-known solutions in the classical setting are apparently not known.

Harmonic functions on discrete structures were introduced by Phillips and Wiener (1923). They solved the Dirichlet problem for harmonic functions on subsets of $\mathbf{Z}^n$ using the minimum of a positive definite quadratic form. Blanc (1939) solved the Dirichlet problem on a circular net in the plane by classical arguments from linear algebra. Duffin (1953) developed a discrete potential theory, including the study of a fundamental solution for the Laplacian in $\mathbf{Z}^3$.

In theoretical physics there has been an interest in a discrete calculus during several years as witnessed, for instance, by Novikov and Dynnikov (1997) and Guo and Wu (2003).

Recently several studies of a discrete Laplace operator have appeared, for instance, that of Chung and Yau (2000), who solve the Dirichlet problem in terms of eigenfunctions for the Laplacian, and that of Kenyon (2002), where the weight functions are symmetric. A variant of the discrete calculus is when the edges of a graph are equipped with a metric, as in the research of Favre and Jonsson (2004) and Baker and Rumely (2004).

2 Subharmonic functions

Let X be an arbitrary set. We shall say that f is a *structural function on X* if f is defined on the Cartesian product $X^2 = X \times X$, has complex values, and is such that the set $\{y \in X; f(x, y) \neq 0\}$ is finite for every $x \in X$. The set of all structural functions is an algebra with addition defined pointwise:

$$(f + g)(x, z) = f(x, z) + g(x, z), \quad (x, z) \in X^2, \tag{2.1}$$

and multiplication defined as

$$(f \diamond g)(x, z) = \sum_{y \in X} f(x, y)g(y, z), \quad (x, z) \in X^2. \tag{2.2}$$

If X is finite, this is just a matrix algebra, but it is convenient to use the functional notation even then. The multiplication generalizes both pointwise multiplication and convolution. Indeed, if $f(x, y) = F(x, y)\delta(x, y)$ and $g(x, y) = G(x, y)\delta(x, y)$ with the Kronecker delta, then $(f \diamond g)(x, z) = F(x)G(x)\delta(x, z)$; if X is an abelian group and $f(x, y) = F(x - y)$, $g(x, y) = G(x - y)$, then

$$(f \diamond g)(x, z) = \sum_y F(x - y)G(y - z) = (F * G)(x - z),$$

the convolution product of F and G.

We shall say that a structural function $\lambda \colon X^2 \to \mathbf{R}$ is a *weight function* if $\lambda \geqslant 0$ and $\sum_y \lambda(x, y) > 0$ for every $x \in X$.

The inequality

$$f(x) \sum_{y \in X} \lambda(x, y) \leqslant \sum_{y \in X} \lambda(x, y) f(y) \tag{2.3}$$

has a sense for all weight functions λ and all functions $f \colon X \to \mathbf{R}$ and even for functions with values in $[-\infty, +\infty[$ if we define $0 \cdot (-\infty) = 0$.

Let us define the *Laplacian* $\Delta f = \Delta_\lambda f$ of f at a point x as

$$\Delta f(x) = \sum_{y \in X} \lambda(x, y)(f(y) - f(x)), \quad x \in X. \tag{2.4}$$

We shall say that $f: X \to \mathbf{R}$ is *subharmonic at* x if $\Delta f(x) \geqslant 0$, and that it is *subharmonic in* X if $\Delta f(x) \geqslant 0$ for all $x \in X$. As usual we shall say that f is *superharmonic* if $-f$ is subharmonic, and *harmonic at* x (*in* X) if $\Delta f(x) = 0$ (for all $x \in X$, respectively). All this depends, of course, on the choice of weight function.

Given a point $x \in X$, the points y such that $\lambda(x, y) > 0$ may be called *neighbors* of x; note that this relation need not be symmetric. We have a directed graph where there is an arrow from x to all its neighbors, and we thus compare the value at x with a weighted mean value over all neighbors of x.

Example 2.1. Extreme examples are the following. If $\lambda(x, y) = \delta(x, y)$, then every function is harmonic; $\Delta_\delta = 0$. If X is finite and $\lambda(x, y) = 1$ for all x, y, then only the constant functions are subharmonic.

Example 2.2. Let $X = \mathbf{Z}$ and $\lambda(x, y) = \delta(x + 1, y)$. Then a function is subharmonic exactly when it is increasing.

Example 2.3. The solutions to a discrete analogue of the heat equation,

$$u(x, s + 1) - u(x, s) = \kappa(u(x - 1, s) - 2u(x, s) + u(x + 1, s))$$

are λ-harmonic if we define $\lambda((x, s), (y, t)) = \kappa$ when $(y, t) = (x \pm 1, s - 1)$, $\lambda((x, s), (y, t)) = 1 - 2\kappa$ when $(y, t) = (x, s - 1)$, and zero otherwise. Here we assume that $0 \leqslant \kappa \leqslant \frac{1}{2}$. Another discrete analogue of the heat equation is

$$u(x, s) - u(x, s - 1) = \kappa(u(x - 1, s) - 2u(x, s) + u(x + 1, s));$$

here we define $\lambda((x, s), (y, t)) = \kappa$ when $(y, t) = (x \pm 1, s)$, $\lambda((x, s), (y, t)) = 1$ when $(y, t) = (x, s - 1)$, and zero otherwise. Here any $\kappa \geqslant 0$ will do. The two equations have very different properties.

A natural interpretation of the Laplacian is in terms of random walks: $\lambda(x, y)$ is then the transition probability from x to y, assuming that $\sum_y \lambda(x, y) = 1$; see, e.g., Chung and Yau (2000). If the weight function λ is symmetric, i.e., $\lambda(x, y) = \lambda(y, x)$, then there is an interpretation of harmonic functions as potentials in an electric circuit. We let $1/\lambda(x, y)$ be the resistance of the link between x and y. Then harmonicity at a point x means exactly that Kirchhoff's law holds: the sum of all outgoing currents from x is equal to the sum of all incoming currents to x. However, for symmetric weight functions the Dirichlet problem as we formulate it is not interesting.

Let $X = \mathbf{Z}^2$ and let $\lambda(x, y) = 1$ if $\|y - x\|_1 = 1$ and zero otherwise. Then a function is harmonic at a point x if and only if its value at x is equal to the arithmetic mean of its values at its four neighbors $(x_1 \pm 1, x_2)$, $(x_1, x_2 \pm 1)$. We shall call such functions $\mathbf{Z}^2$-*harmonic* and define the $\mathbf{Z}^2$-Laplacian as

$$\Delta_{\mathbf{Z}^2} u(x) = u(x_1 + 1, x_2) + u(x_1 - 1, x_2) + u(x_1, x_2 + 1) + u(x_1, x_2 - 1) - 4u(x).$$

We shall compare this with the classical Laplacian,

$$\Delta_{\mathbf{R}^2} u = \frac{\partial^2 u}{\partial x^2} + \frac{\partial^2 u}{\partial y^2}.$$

Example 2.4. A second-degree polynomial $u(x, y) = ax^2 + 2bxy + cy^2$, where $(x, y) \in \mathbf{R}^2$ or $(x, y) \in \mathbf{Z}^2$, has Laplacians $\Delta_{\mathbf{R}^2} u = 2(a + c)$, $\Delta_{\mathbf{Z}^2} u = 2(a + c)$, respectively. It is therefore harmonic simultaneously in the two settings: if and only if $a + c = 0$.

Example 2.5. The exponential functions behave differently in the two cases. An exponential function $\mathbf{R}^2 \ni (x, y) \mapsto e^{\alpha x + \beta y}$, where α and β are complex numbers, is harmonic in the classical sense if and only if $\beta = \pm i\alpha$. In fact, the $\mathbf{R}^2$-Laplacian is

$$\Delta_{\mathbf{R}^2} e^{\alpha x + \beta y} = (\alpha^2 + \beta^2)e^{\alpha x + \beta y}.$$

In particular, we have the functions $\cos(\beta x)e^{-\beta y}$, $\beta \in \mathbf{R}$, which are harmonic and tend rapidly to zero in the upper half plane if β is a large positive number. By way of contrast, the exponential function $\mathbf{Z}^2 \ni (x, y) \mapsto e^{\alpha x + \beta y}$ is $\mathbf{Z}^2$-harmonic if and only if $\cosh \alpha + \cosh \beta = 2$; its $\mathbf{Z}^2$-Laplacian is

$$\Delta_{\mathbf{Z}^2} e^{\alpha x + \beta y} = (2 \cosh \alpha + 2 \cosh \beta - 4)e^{\alpha x + \beta y}.$$

The function $h(x, y) = \cos(\alpha x)e^{-\beta y}$, where $\alpha, \beta \in \mathbf{C}$, is harmonic if and only if $\cos \alpha + \cosh \beta = 2$, thus if and only if

$$e^{-\beta} = 2 - \cos \alpha \pm \sqrt{(2 - \cos \alpha)^2 - 1}.$$

If we take α real here and choose the minus sign, we note that $e^{-\beta}$ satisfies $3 - \sqrt{8} \leqslant e^{-\beta} \leqslant 1$, so that the functions tend to zero, but not as rapidly as in the real case. Choosing $\alpha = \pi$ we get $e^{-\beta} = 3 - \sqrt{8}$ and

$$h(x, y) = \cos(\pi x)e^{-\beta y} = (-1)^x \left(3 - \sqrt{8}\right)^y, \quad (x, y) \in \mathbf{Z}^2,$$

with the fastest possible decay as $y \to +\infty$ for this class of functions. Therefore, $\mathbf{R}^2$-harmonic functions cannot be well approximated by $\mathbf{Z}^2$-harmonic ones. We can get a faster decay as $y \to +\infty$ only if we allow growth in the x direction: $h(x, y) = e^{i\alpha x - \beta y}$ is harmonic for large positive β and a suitable α, but then α cannot be real, implying that h is unbounded on the real axis.

We shall say that a weight function λ is *normalized* if $\sum_{y \in X} \lambda(x, y) = 1$ for all $x \in X$. To any weight function λ we associate a normalized weight function

$$\lambda'(x, y) = \frac{\lambda(x, y)}{\sum_{z \in X} \lambda(x, z)}, \quad (x, y) \in X^2. \tag{2.5}$$

If λ is normalized, the Laplacian may be written

$$\Delta f(x) = \sum_{y \in X} (\lambda(x, y) - \delta(x, y))f(y), \quad x \in X.$$

We note that $I + \Delta$ is an increasing operator: $f \leqslant g$ implies $(I + \Delta)f \leqslant (I + \Delta)g$.

Proposition 2.6. *Assume that u is subharmonic with respect to two weight functions* λ *and* μ. *Let* $c\colon X \to \mathbf{R}$ *be a function with nonnegative values. Then u is subharmonic with respect to* λ' *(defined by (2.5)),* $\lambda(x, y) + c(x)\mu(x, y)$, *and* $\lambda \diamond \mu'$. *Moreover, it is subharmonic with respect to* $\lambda(x, y) + g(x)\delta(x, y)$ *for any real-valued function g such that* $g(x) + \lambda(x, x) \geqslant 0$ *and* $g(x) + \sum_{y \in X} \lambda(x, y) > 0$ *for all* $x \in X$. *(We may thus choose* $g(x) = -\lambda(x, x)$ *except when* $\lambda(x, y) = 0$ *for all* $y \neq x$.)*

The proofs are straightforward. We note that the Laplacian of $\lambda \diamond \mu$ is given by

$$\Delta_{\lambda \diamond \mu} u(x) = \sum_{y} \lambda(x, y) \Delta_{\mu} u(y) + \Delta_{\lambda} u(x) \tag{2.6}$$

if μ is normalized. If both λ and μ are normalized, this can be written succinctly as

$$I + \Delta_{\lambda \diamond \mu} = (I + \Delta_{\lambda}) \circ (I + \Delta_{\mu}). \tag{2.7}$$

Example 2.7. Let X be the infinite strip $\{x \in \mathbf{Z}^2; x_2 = 0, 1\}$ with only two pixels in the vertical direction. We define the weight function for $(x, y) \in X$ by $\lambda(x, y) = 1$ when $y = (x_1 \pm 1, x_2)$; $\lambda(x, y) = \kappa$ when $y = (x_1, 1 - x_2)$ and zero otherwise. Here $\kappa \geqslant 0$ is a kind of coupling constant. Thus

$$\Delta u(x) = u(x_1 - 1, x_2) + \kappa u(x_1, 1 - x_2) + u(x_1 + 1, x_2) - (2 + \kappa)u(x), \quad x \in X.$$

Let $\tau \geqslant 1$ be the largest solution to the equation $\tau^2 - 2(\kappa + 1)\tau + 1 = 0$ and put $\gamma = \log \tau$. The function

$$U(x) = (2x_2 - 1)\tau^{x_1} = (2x_2 - 1)e^{\gamma x_1}, \quad x \in X, \tag{2.8}$$

is harmonic.

By combining U and its reflection in the x_1 variable and then performing a translation we get further examples for all choices of real constants a and C,

$$U_0(x) = (2x_2 - 1) \cosh \gamma (x_1 - a) + C(x_1 - a), \quad x \in X, \tag{2.9}$$

and

$$U_1(x) = (2x_2 - 1) \sinh \gamma (x_1 - a) + C(x_1 - a), \quad x \in X, \tag{2.10}$$

with odd symmetry in the line $x_1 = a$. This is an infinitely long finger, and the solution will serve as a building block for the case of a finitely long finger.

3 The Dirichlet problem on finite sets

Given a weight function λ on a set X we define the *boundary* of X, denoted by ∂X, as the set of all points $x \in X$ such that $\lambda(x, y) = 0$ for all $y \neq x$. Its complement $X^\circ = X \smallsetminus \partial X$ is the *interior* of X. We note that a point x has a neighbor different from x if and only if x is in the interior.

Given a point $a \in X$, we define $N^0(a) = \{a\}$ and then inductively

$$N^{k+1}(a) = \{y; \lambda(x, y) > 0 \text{ for some } x \in N^k(a)\}, \quad k \in \mathbf{N}.$$

The union of all the $N^k(a)$ will be called the λ-*component* of a and be denoted by $C(a)$. Clearly $C(a) = \{a\}$ if and only if a is a boundary point.

We shall say that X is *boundary connected* if $C(a)$ intersects ∂X for every $a \in X^\circ$. (In particular ∂X is nonempty if X is boundary connected and nonempty.) We shall say that X is *connected* if $C(a) = X$ for all $a \in X^\circ$. (Here ∂X may be empty.)

Proposition 3.1. *If X is finite and boundary connected, then $\sup_X u = \sup_{\partial X} u$ for all subharmonic functions u on X. The converse holds.*

Proof. Take a point $a \in X$ such that $u(a) = \sup_X u = A$. Then u must take the value A also at all points y such that $\lambda(a, y) > 0$, in other words in the set $N^1(a)$. Continuing this argument, we see that $u = A$ at all points in $N^k(a)$ and so in their union $C(a)$. By hypothesis $C(a)$ contains a point in ∂X. We are done.

For the converse we note that the characteristic function $\chi_{C(a)}$ is subharmonic in X and if $C(a)$ is contained in X°, then it is zero on the boundary but takes the value one at a.   □

Example 3.2. If X is boundary connected and u is harmonic and constant on ∂X (in particular if ∂X consists of just one point), then u is constant in all of X. It might be surprising that one point suffices to keep u constant. If $X = [0, m]_{\mathbf{Z}}$ with $\partial X = \{0\}$ and $\lambda(x, y) = 1$ when $x \in X^\circ$ and $y \in X$ with $y = x \pm 1$ and zero otherwise, then this result applies. (The harmonicity at $x = m$ forces $u(m - 1)$ to be equal to $u(m)$.) If, on the other hand, X is the infinite interval $[0, +\infty[_{\mathbf{Z}}$, then all functions $u(x) = ax + b$ are harmonic.

Proposition 3.3. *If X is connected (finite or infinite), then a subharmonic function which attains its supremum at an interior point must be constant. The converse holds.*

Proof. Assume that there exists a point $a \in X$ such that $u(a) = \sup_X u = A$. Then, just as in the previous proof, u must take the value A also at all points in $C(a)$. But by assumption this is all of X.

For the converse we note again that $\chi_{C(a)}$ is subharmonic. If $C(a) \neq X$ it is not constant.   □

We shall now take a look at the Dirichlet problem in finite sets X. This is about finding u such that $\Delta u = f$ in X and $u = g$ on ∂X for given functions f and g. Since a function is always harmonic at a boundary point, we must have $f = 0$ on ∂X; equivalently, we let f be given in X° and require only that $\Delta u = f$ in the interior X°.

From linear algebra we know that a system of equations $\Delta u(x) = f(x)$ has a solution if f satisfies as many independent linear conditions as there are independent functions ρ satisfying

$$\sum_{x \in X} \rho(x)(\lambda(x, y) - \delta(x, y)) = 0, \quad y \in X. \tag{3.1}$$

More precisely, if (3.1) holds, then we must have $\sum_x \rho(x) f(x) = 0$. (We assume λ to be normalized here.) We can always take $\rho(x) = \delta(a, x)$ if a is a boundary point. This gives the condition that $f(a)$ must vanish as we already noted. But there may be other such functions ρ. For instance, if λ is symmetric, $\lambda(x, y) = \lambda(y, x)$, then we may take ρ equal to 1 identically, and we conclude that the equation $\Delta u = f$ can be solved only if $\sum_{x \in X} f(x) = 0$. In this case the set X is not boundary connected: $C(a)$ is contained in X° for all $a \in X^\circ$.

Even though the equation $\Delta u(x) = f$ is a finite system of linear equations it is convenient to express conditions for solvability in terms of subsolutions, just as in the real case. The resulting theorem is more general than those of Phillips and Wiener (1923) and Blanc (1939).

Theorem 3.4. *Let X be a finite set and λ a weight function on X. Assume that X is boundary connected. Let two functions be given: $f \geqslant 0$ in X° and g on ∂X. Assume also that the Dirichlet problem has a subsolution, i.e., that there exists a function w such that $\Delta w \geqslant f$ in X° and $w \leqslant g$ on ∂X. Then the Dirichlet problem $\Delta u = f$ in X°, $u = g$ on ∂X, has a unique solution.*

We note that if f is identically zero, then there is always a subsolution: we may take w as a constant not exceeding $\inf_{x \in \partial X} g(x)$. Also, if X is a subset of $\mathbf{Z}^2$ and we define $\lambda(x, y) = 1$ when $\|y - x\|_1 = 1$ and the five points y with $\|y - x\|_1 \leqslant 1$ all lie in X, and $\lambda(x, y) = \delta(x, y)$ otherwise, then there is always a subsolution: we may take $w(x) = c_1 \|x\|_2^2 - c_2$ for sufficiently large constants c_1 and c_2.

As already noted before the statement of the theorem, there are cases when there is no solution (hence no subsolution).

Proposition 3.5. *Under the hypotheses in Theorem 3.4, the unique solution to the Dirichlet problem is given as the supremum of all subsolutions: define*

$$u(x) = \sup_w (w(x); \Delta w \geqslant f \text{ in } X^\circ \text{ and } w \leqslant g \text{ on } \partial X).$$

Then u solves the Dirichlet problem: $\Delta u = f$ in X°, $u = g$ on ∂X.

Proof. We define an operator T by the equations $T(w) = w + \Delta w - f$ in X°, $T(w) = g$ on ∂X. We assume here that λ is normalized, i.e., that $\sum_y \lambda(x, y) = 1$, which is no restriction. If w is a subsolution, then $T(w) \geqslant w$. Then also $T(w)$ is a subsolution in view of the fact that $I + \Delta$ is increasing: $(I + \Delta)T(w) \geqslant (I + \Delta)w = T(w) + f$, which implies that $\Delta T(w) \geqslant f$. The boundary condition is of course satisfied.

Let now u be the supremum of all subsolutions. Since there exists at least one subsolution, $u > -\infty$. On the other hand $u < +\infty$, for a subsolution can never exceed $\sup_{x \in \partial X} g(x)$ in view of Proposition 3.1. Thus u has real values. We claim that u itself is a subsolution. Let w be any subsolution. Then $u \geqslant w$ and, since $I + \Delta$ is increasing, also $(I + \Delta)u \geqslant (I + \Delta)w \geqslant w + f$. Taking the supremum over all w we get $(I + \Delta)u \geqslant \sup_w w + f = u + f$, which shows that u is a subsolution.

Now also $T(u) = u + \Delta u - f$ is a subsolution; hence $T(u) \leqslant u$. But we already proved that $T(u) \geqslant u$. Therefore, $T(u) = u$, which implies that $\Delta u = f$. $\qquad\square$

Proof of Theorem 3.4. Only uniqueness remains to be considered. Let us assume that we have two solutions u and v. Then their difference $u - v$ is harmonic and Proposition 3.1 shows that $\sup_X (u - v) = \sup_{\partial X} (u - v)$. But $u - v = g - g = 0$ on the boundary. Hence $u - v \leqslant 0$ in X; on interchanging the roles of u and v we obtain $v - u \leqslant 0$ in X. □

Proposition 3.6. *There is a comparison principle: if X is finite and boundary connected, and if u and v satisfy $\Delta u \geqslant \Delta v$ in X° and $u \leqslant v$ on ∂X, then $u \leqslant v$ in all of X.*

Proof. The function $w = u - v$ is subharmonic and satisfies $w \leqslant 0$ on the boundary. Hence also in the interior by Proposition 3.1. □

Remark 3.7. We can get the solution as the limit of a sequence. We may start with $u_0 = w$, any subsolution, and then define $u_{j+1} = T(u_j)$, $j \in \mathbf{N}$. The sequence (u_j) is increasing as we have seen, and all the u_j are subsolutions. It is bounded from above, for in view of Proposition 3.1 it can never exceed $\sup_{x \in \partial X} g(x)$. Hence the sequence admits a limit u. If we pass to the limit in the definition $u_{j+1} = u_j + \Delta u_j - f$ we obtain $\Delta u = f$. The boundary condition $u = g$ on ∂X is preserved. The convergence may be very slow, however. Better start with a very good approximation.

Example 3.8. Let X be the cube $\{0, 1\}^n$ in $\mathbf{Z}^n$, and let the boundary consist of the two points 0 and $(1, 1, \ldots, 1)$. Let the weight function be $\lambda(x, y) = 1$ if x is an interior point and $\|y - x\|_1 = 1$, and zero otherwise. Then the solution to the Dirichlet problem with zero Laplacian and boundary values $u(0) = 0$, $u(1, 1, \ldots, 1) = 1$ is

$$u(x) = C \sum_{j=0}^{k-1} \binom{n-1}{j}^{-1}, \quad \text{where } k = \sum_{j=1}^{n} x_j \quad \text{and} \quad C = \left[\sum_{j=0}^{n-1} \binom{n-1}{j}^{-1} \right]^{-1}.$$

Here

$$\binom{n-1}{j}^{-1} = \frac{j!(n-1-j)!}{(n-1)!}$$

is the inverse of the binomial coefficient. For example, in the six-dimensional cube we have $u(0, 0, 0, 0, 0, 1) = \frac{5}{13}$ and $u(0, 0, 0, 0, 1, 1) = \frac{6}{13}$, the other values being easily obtained from these two and the symmetry.

Example 3.9. Let us look at a narrow rectangle in $\mathbf{Z}^2$ with boundary at two vertices: $X_m = \{x \in \mathbf{Z}^2; x_1 = 0, \ldots, m, x_2 = 0, 1\}$. Let the boundary be $\partial X_m = \{(0, 0), (m, 0)\}$ and let the weight function be $\lambda(x, y) = \delta(x, y)$ if $x \in \partial X_m$; $\lambda(x, y) = 1$ if $y = (x_1 \pm 1, x_2)$ and $x, y \in X_m$; and $\lambda(x, y) = \kappa$ when $y = (x_1, 1 - x_2)$, $x_1 = 1, \ldots, m - 1$ or $(x, y) = ((0, 1), (0, 0))$ or $(x, y) = ((m, 1), (m, 0))$. Then the function U_1 defined by (2.10) is harmonic at all points in X_m except at the two vertices $(0, 1)$ and $(m, 1)$ where we have changed the weight function compared with the infinite strip in Example 2.7. This is true for all choices of the constants a and C. We now choose $a = \frac{1}{2}m$ to get odd symmetry in the line

$x_1 = \frac{1}{2}m$. If $\kappa > 0$, there is a unique constant C such that U_1 is harmonic at $(0, 1)$ (and by the symmetry also at $(m, 1)$). This value of C is

$$C = -\frac{1}{2}(\tau - 1)(\tau^{m/2} + \tau^{-m/2-1}) = -2\sinh\left(\frac{1}{2}\gamma\right)\cosh\frac{1}{2}\gamma(m+1),$$

and the boundary values are

$$U_1((0, 0)) = -U_1((m, 0)) = \sinh\frac{1}{2}\gamma m + m\sinh\left(\frac{1}{2}\gamma\right)\cosh\frac{1}{2}\gamma(m+1),$$

where τ and $\gamma = \log\tau$ are defined as in Example 2.7. After a normalization we get the solution $u = U_1/U_1((m, 0))$ to the Dirichlet problem $\Delta u = 0$ in X_m, $u = -1, 1$ at the two boundary points $(0, 0)$ and $(m, 0)$, respectively. How much does it deviate from the affine function $v(x) = -1 + 2x_1/m$, which is harmonic at all points $x \in X_m$ except at the two vertices $(0, 1)$ and $(m, 1)$?

4 Clustering of values

We shall now study a phenomenon which can be expressed as clustering of values of harmonic functions.

On a long narrow object, clustering can be severe as shown by the following example. But by relaxing the coupling between strategically chosen parts of the structure, the clusters can be resolved.

Example 4.1. Let us consider $X_m = \{x \in \mathbf{Z}^2; 0 \leqslant x_1 \leqslant m, 0 \leqslant x_2 \leqslant 1\}$, but now with the boundary $\partial X = \{(0, 0), (0, 1)\}$. We prescribe the boundary values $g(0, 0) = -1$, $g(0, 1) = 1$. The weight function shall be $\lambda(x, y) = 1$ if x is an interior point and $y = (x_1 \pm 1, x_2) \in X_m$; $\lambda(x, y) = \kappa \geqslant 0$ if x is an interior point and $y = (x_1, 1 - x_2)$; otherwise $\lambda(x, y) = 0$. We would like to illustrate what happens if we vary the weight by varying the coupling constant κ. When $\kappa = 0$ there is no coupling between the points $(x_1, 0)$ and $(x_1, 1)$ and the solution to the Dirichlet problem is $u(x) = 2x_2 - 1$.

In view of the symmetry, the solution to the Dirichlet problem with zero Laplacian must satisfy $u(x_1, 0) = -u(x_1, 1)$. We define u as AU_0 in (2.9), where $C = 0$ and where a and A are constants to be determined. This function is harmonic everywhere in X_m except possibly when $x_1 = m$. We determine a so that it becomes harmonic also at the two points $(m, 0)$, $(m, 1)$; this happens if and only if $a = m + \frac{1}{2}$. To give the function the value 1 at $(0, 1)$, we define

$$A = \frac{1}{U_0((0, 1))} = \frac{1}{\cosh\gamma(m + \frac{1}{2})} = \frac{2\tau^{-m-1/2}}{1 + \tau^{-2m-1}}.$$

The function $v(x) = (2x_2 - 1)\tau^{-x_1}$ is harmonic in the infinite set $X = \mathbf{N} \times \{0, 1\}$. Its restriction to $0 \leqslant x_1 \leqslant m$ is close to being harmonic in X. In fact, it is harmonic at every point $x \in X$ with $x_1 < m$; at the point $(m, 1)$ it is subharmonic, and at the point $(m, 0)$ superharmonic:

$$\Delta v(m, 1) = -\Delta v(m, 0)$$
$$= v(m - 1, 1) + \kappa v(m, 0) - (1 + \kappa)v(m, 1) = (1 - 1/\tau)\tau^{-m},$$

which is a small number for large m. So it is natural to guess that v is a good approximation of u.[1] In fact,

$$u(j, 1) = a_j = \tau^{-j} + \frac{\tau^{-2m-1}(\tau^j - \tau^{-j})}{1 + \tau^{-2m-1}}.$$

The relative deviation of the true solution u from the comparison function v is

$$\frac{u(j, 1) - v(j, 1)}{v(j, 1)} = \frac{a_j - \tau^{-j}}{\tau^{-j}} = \frac{\tau^{-2m-1}(\tau^{2j} - 1)}{1 + \tau^{-2m-1}} \leqslant \tau^{2(j-m)-1} \leqslant \frac{1}{\tau}.$$

It is clear that the solution u to the Dirichlet problem decays exponentially and that the values lie extremely close to each other if, for instance, $\kappa = 1$ and $m = 100$. Then $\tau = 2 + \sqrt{3} \approx 3.73$ and $\tau^{-100} \approx 6 \cdot 10^{-58}$. If κ is small, the decay is slower: $\kappa = .01$ yields $\tau \approx 1.15$ and $\tau^{-100} \approx 7 \cdot 10^{-7}$. In this simple example it is thus possible to dissolve the clusters by relaxing the coupling.

Example 4.2. Let us look at an example in three dimensions. Let Y_m be the set of points $x = (x_1, x_2, x_3)$ in $\mathbf{Z}^3$ satisfying $0 \leqslant x_1 \leqslant m, 0 \leqslant x_2, x_3 \leqslant 1$. Its boundary shall be $\{(0, 0, 0), (0, 1, 0), (0, 0, 1), (0, 1, 1)\}$, the four points in Y_m with $x_1 = 0$. For an interior point x we take $\lambda(x, y) = 1$ if $y \in Y_m$, $\|y - x\|_1 = 1$, $y_1 = x_1$; we take $\lambda(x, y) = \kappa$ if $y \in Y_m$, $\|y - x\|_1 = 1$, $y_2 = 1 - x_2$; and zero otherwise. We define boundary values $g(0, 0, 0) = g(0, 0, 1) = -1$, $g(0, 1, 0) = g(0, 1, 1) = 1$. Then the solution to the Dirichlet problem with zero Laplacian satisfies $-u(x_1, 0, x_3) = u(x_1, 1, x_3) = a_{x_1}$, where the sequence (a_j) is the same as in the previous example. Thus the phenomenon of rapid exponential decay occurs also in three dimensions when there is a narrow finger; in this case consisting of four voxels at each level $x_1 = $ constant. Of course this finger can be a part of a larger set provided the latter is symmetric around the plane $x_2 = \frac{1}{2}$.

5 The Dirichlet problem on infinite sets

Let us take a look at a Dirichlet problem on infinite sets.

Theorem 5.1. *Let X be a finite or infinite set and λ a weight function on X. We shall consider a Dirichlet problem,*

---

[1] One could perhaps believe that it would be enough to change v at the points $(m, 1)$ and $(m, 0)$ so that it becomes subharmonic or superharmonic in all of X (to be able to use the comparison principle). But this is not possible! One must change the value at several points.

$$\Delta u = f \quad \text{in } X^{\circ}, \quad u = g \quad \text{on } Y,$$

where Y is a subset of X, not necessarily contained in the boundary, and where f and g have real values. Assume that X can be written as a disjoint union $X = \bigcup_0^N X_j$ *or* $X = \bigcup_0^{\infty} X_j$ *with* $X_0 \subset Y$. *In the finite case we assume that* X_N *is contained in the boundary. Assume, moreover, that for every* $y \in X_{j+1}$ *either y belongs to Y or there is a unique* $x \in X_j$ *such that* $\lambda(x, y) > 0$, *that*

$$\{z; \lambda(x, z) > 0\} \subset X_0 \cup \cdots \cup X_j \cup \{y\},$$

and that every $x \in X_j$ *occurs in this way. (There is thus a bijection between* $X_{j+1} \setminus Y$ *and* X_j.) *Then the Dirichlet problem has a unique solution.*

Proof. We define first $u = g$ in X_0. Then if u has already been defined in

$$X_0 \cup \cdots \cup X_j$$

satisfying $\Delta u = f$ in $X_0 \cup \cdots \cup X_{j-1}$ (the empty set if $j = 0$) and $u = g$ in $Y \cap (X_0 \cup \cdots \cup X_j)$, we define u at a point $y \in X_{j+1}$ so that it satisfies $\Delta u = f$ at every point in X_j. Indeed, if $y \in Y$ we just define $u(y) = g(y)$; if not, we need to solve the equation

$$\lambda(x, y)(u(y) - u(x)) + \sum_{z \neq y} \lambda(x, z)(u(z) - u(x)) = f(x),$$

where x is uniquely determined from y. The equation has only one unknown $u(y)$; the coefficient in front of it is nonzero by hypothesis and all the z that occur in the sum belong to $X_0 \cup \cdots \cup X_j$. We do this for every $y \in X_{j+1}$. The extended function is now defined in $X_0 \cup \cdots \cup X_{j+1}$ and satisfies $\Delta u = f$ in $X_0 \cup \cdots \cup X_j$, so the induction step is completed, and the induction goes on in the infinite case. In the finite case it stops, and we note that we need not require that u be harmonic in X_N since this is automatic in view of our assumption that X_N is contained in the boundary. $\square$

Example 5.2. Let $X = \{x \in \mathbf{Z}^2; x_2 \geqslant 0\}$ with $\lambda(x, y) = 1$ if $\|y - x\|_1 = 1$, $x, y \in X$, and zero otherwise. Let $Y = \{x \in X; x_2 = 0\}$. (Note that Y is not contained in the boundary; the latter is in fact empty.) Here we can define $X_j = \{x \in X; x_2 = j\}$, $j \in \mathbf{N}$. The theorem shows that the Dirichlet problem $\Delta u = f$ in X, $u = g$ on Y has a unique solution. This problem is equivalent to constructing a solution (with another weight function) in $\mathbf{Z}^2$ which is symmetric in the second variable.

Example 5.3. Let $X = \{x \in \mathbf{Z}^2; x_1 \geqslant |x_2|\}$ with $\lambda(x, y) = 1$ if $\|y - x\|_1 = 1$, $x, y \in X$, and zero otherwise. Let $Y = \{x \in X; x_1 = |x_2|\}$. (The boundary is empty also in this case.) We define now $X_j = \{x \in X; x_1 = j\}$. The theorem says that there is a unique solution to the Dirichlet problem. This problem is equivalent to solving the Dirichlet problem with a function u which is symmetric in the sense that $u(x_1, x_2) = u(x_2, x_1)$ and $u(x_1, x_2) = u(-x_2, -x_1)$ and prescribed on the diagonals.

6 The Poisson kernel

The Poisson kernel in two real variables is

$$P_y(x) = \frac{1}{\pi} \frac{y}{x^2 + y^2}, \quad (x, y) \in \mathbf{R} \times \mathbf{R}^+.$$

It satisfies the convolution equations

$$P_y * P_{y'} = P_{y+y'}, \quad y, y' \in \mathbf{R}^+,$$

where the convolution, denoted by $*$, is in the x-variable only. Its Fourier transform with respect to x is

$$\widehat{P}_y(\xi) = \int_{\mathbf{R}} P_y(x) e^{-i\xi x} dx = e^{-y|\xi|}, \quad y \in \mathbf{R}^+, \quad \xi \in \mathbf{R},$$

from which the convolution property is also evident; $\widehat{P}_y = \widehat{P}_1^y$ for all $y > 0$. If we define P_0 as the Dirac measure at the origin, then the convolution equation extends to all $y \geq 0$.

The Poisson kernel in $\mathbf{Z}^2$ shall be defined in a similar way. It is a function $Q_y(x)$ of $(x, y) \in \mathbf{Z} \times \mathbf{N}$ which is $\mathbf{Z}^2$-harmonic at every point (x, y) with $y \geq 1$ and satisfies $Q_y * Q_{y'} = Q_{y+y'}$. Its Fourier transform

$$\widehat{Q}_y(\xi) = \sum_{x \in \mathbf{Z}} Q_y(x) e^{-ix\xi}, \quad \xi \in \mathbf{R},$$

satisfies $\widehat{Q}_y = \widehat{Q}_1^y$, $y \in \mathbf{N}$.

If we use the harmonicity of Q at points (x, y) with $y = 1$ and the condition that $Q_0 = Q_1^0 = \delta$, $Q_2 = Q_1 * Q_1$, then $\widehat{Q}_1$ must satisfy an equation of the second degree,

$$\widehat{Q}_1(\xi)^2 + e^{-i\xi} \widehat{Q}_1(\xi) + e^{i\xi} \widehat{Q}_1(\xi) + 1 = 4\widehat{Q}_1(\xi),$$

so that

$$\widehat{Q}_1(\xi) = 2 - \cos \xi \pm \sqrt{(2 - \cos \xi)^2 - 1}, \quad \xi \in \mathbf{R},$$

where we may choose the sign $\pm$ for each $\xi \in \mathbf{R}$ in a measurable way. If we always choose the negative sign, we get a decreasing function of y,

$$\widehat{Q}_y(\xi) = \left(2 - \cos \xi - \sqrt{(2 - \cos \xi)^2 - 1}\right)^y, \quad y \in \mathbf{N}, \quad \xi \in \mathbf{R}.$$

We can state that Q_y is the inverse Fourier transform of $\widehat{Q}_y = \widehat{Q}_1^y$:

$$Q_y(x) = \frac{1}{2\pi} \int_{-\pi}^{\pi} \widehat{Q}_1(\xi)^y e^{i\xi x} d\xi, \quad x \in \mathbf{Z}, \quad y \in \mathbf{N}.$$

So there is an explicit formula for $\widehat{Q}_y$ but hardly for Q_y. However, $Q_y(x)$ is very close to $P_y(x)$.

7 Fundamental solutions

In two real variables a well-known fundamental solution for the Laplacian is

$$E(x, y) = \frac{1}{4\pi} \log(x^2 + y^2), \quad (x, y) \in \mathbf{R}^2.$$

We would like to determine a fundamental solution of the $\mathbf{Z}^2$-Laplacian.

Lemma 7.1. *Let a be a nonnegative number and define*

$$E_a(x, y) = \log(a + x^2 + y^2) = \log(a + r^2), \quad (x, y) \in \mathbf{Z}^2,$$

where $r = \sqrt{x^2 + y^2}$. It is subharmonic in all of $\mathbf{Z}^2$ if and only if $a \geqslant \frac{1}{2}$.

Proof. We find that $\Delta E_a = \log \frac{N}{D}$, where

$$N = r^8 + 4ar^6 + (6a^2 + 4a - 2)r^4 + (4a^3 + 8a^2 + 4a)r^2 + (a + 1)^4 + 16x^2y^2$$

and

$$D = (a + r^2)^4 = r^8 + 4ar^6 + 6a^2r^4 + 4a^3r^2 + a^4.$$

Now this function is nonnegative if and only if $N/D \geqslant 1$; equivalently, $N - D \geqslant 0$. We get

$$N - D = (4a - 2)r^4 + (8a^2 + 4a)r^2 + 4a^3 + 6a^2 + 4a + 1 + 16x^2y^2,$$

from which the result is obvious.   □

We see that for all a, the function is very close to being harmonic far away from the origin: $\Delta E_a = O(r^{-4})$ as $r \to +\infty$. When $a = 0$ the function is superharmonic in sectors of openings $45°$ around the x-axis and y-axis far away from the origin, and subharmonic in sectors of openings $45°$ around the diagonals far away from the origin.

The fundamental solution E satisfies the differential equation $\Delta E = \delta$. On taking the Fourier transform in the Schwartz space $\mathscr{S}'(\mathbf{R}^2)$, we get

$$-(\xi^2 + \eta^2)\widehat{E}(\xi, \eta) = 1,$$

so that $\widehat{E}$ is a tempered distribution defined in $\mathbf{R}^2$ as an extension of the homogeneous function $-(\xi^2 + \eta^2)^{-1}$ defined in $\mathbf{R}^2 \setminus \{0\}$. One such extension is the pseudofunction $-\mathrm{pf}(\xi^2 + \eta^2)^{-1}$ defined as the finite part of an integral.

Similarly, a fundamental solution for the $\mathbf{Z}^2$-Laplacian shall satisfy $\Delta_{\mathbf{Z}^2} F = \delta$, which implies that its Fourier transform, defined as a distribution in $\mathbf{R}^2$, or rather in $(\mathbf{R} \bmod 2\pi)^2$, shall satisfy

$$(2 \cos \xi + 2 \cos \eta - 4)\widehat{F}(\xi, \eta) = 1.$$

The inverse of the factor in front of $\widehat{F}$ has the same kind of singularity at the origin as $-(\xi^2 + \eta^2)^{-1}$. Therefore, $\widehat{F}$ can be defined by the finite part of an integral of

$(2\cos\xi + 2\cos\eta - 4)^{-1}$. The fundamental solution F is its inverse Fourier transform. Can one find an explicit formula for F?

For the corresponding problem in $\mathbf{Z}^3$, Duffin (1953, p. 239) determined the first terms in the asymptotic development as $\|x\|_2 \to +\infty$ to be

$$F(x) = \frac{1}{4\pi\|x\|_2} + \frac{1}{32\pi\|x\|_2^3}\left[-3 + \frac{5(x_1^4 + x_2^4 + x_3^4)}{\|x\|_2^4}\right] + O(\|x\|_2^{-5}), \quad x \in \mathbf{Z}^3.$$

Burkhardt (1997, p. 1159) proved that there exist asymptotic developments to any order and gave formulas that allow us to calculate them; he gave explicit formulas for the terms of order -5 and -7.

References

M. Baker and R. Rumely (2004), Harmonic analysis on metrized graphs, manuscript; available online from arxiv.org/abs/math/0407427.

C. Blanc (1939–1940), Une interprétation élémentaire des théorèmes fondamentaux de M. Nevanlinna, *Comm. Math. Helv.*, **12**, 153–163.

C. Brechbühler, G. Gerig, and O. Kübler (1995), Parametrization of closed surfaces for 3-D shape description, *Comput. Vision Image Understanding*, **6**-2, 154–170.

R. H. Burkhardt (1997), Asymptotic expansion of the free-space Green's function for the discrete 3-D Poisson equation, *SIAM J. Sci. Comput.*, **18**, 1142–1162.

R. J. Duffin (1953), Discrete potential theory, *Duke Math. J.*, **20**, 233–251.

F. Chung and S.-T. Yau (2000), Discrete Green's functions, *J. Combin. Theory Ser.* A, **91**, 191–214.

C. Favre and M. Jonsson (2004), *The Valuative Tree*, Lecture Notes in Mathematics 1853, Springer-Verlag, New York.

J. Ferrand (1944), Fonctions préharmoniques et fonctions préholomorphes, *Bull. Sci. Math.*, **68**, 152–180.

H.-Y. Guo and K. Wu (2003), On variations in discrete mechanics and field theory, *J. Math. Phys.*, **44**, 5978–6004.

R. Kenyon (2002), The Laplacian and Dirac operators on critical planar graphs, *Invent. Math.*, **150**, 409–439.

S. P. Novikov and I. A. Dynnikov (1997), Discrete spectral symmetries of small-dimensional differential operators and difference operators on regular lattices and two-dimensional manifolds, *Uspekhi Mat. Nauk*, **52**, 1058–1116 (in Russian); *Russian Math. Surveys*, **52** (1997), 1057–1116 (in English).

H. B. Phillips and N. Wiener (1923), Nets and the Dirichlet problem, *J. Math. Phys.*, **2**, 105–124; also available in *Norbert Wiener: Collected Works*, Vol. I, MIT Press, Cambridge, MA, London, 1976, 333–354.

O. Weistrand (2004), Shape approximation of digital surfaces homeomorphic to a sphere, manuscript, Uppsala University, Uppsala, Sweden.

Nearly Hyperbolic Varieties and Phragmén–Lindelöf Conditions

R. W. Braun[1], R. Meise[1], and B. A. Taylor[2]

[1] Mathematisches Institut
Heinrich-Heine-Universität
Universitätsstraße 1
40225 Düsseldorf
Germany
ruediger.braun@uni-duesseldorf.de
meise@math.uni-duesseldorf.de

[2] Department of Mathematics
University of Michigan
Ann Arbor, MI 48109
USA
taylor@umich.edu

Dedicated to Professor Carlos Berenstein in celebration of his 60th birthday.

Abstract. By a classical estimate in function theory, each subharmonic function $u(z)$ on the unit disk that is bounded above by 1 and bounded by 0 on the real axis must satisfy a bound of the form $u(z) \leq A|\operatorname{Im} z|$ on smaller subdisks. When an analogous estimate holds for the plurisubharmonic functions in a neighborhood of a real point ξ in an analytic variety, the variety is said to satisfy the local Phragmén–Lindelöf condition at ξ. Interest in such conditions originated from a theorem of Hörmander who showed that the surjective constant coefficient linear partial differential operators on the space of real analytic functions on $\mathbb{R}^n$ are characterized in terms of these conditions. We give a new geometric condition on a local variety that is necessary in order that the local Phragmén–Lindelöf condition holds, and is sufficient in the case of varieties of dimension 1 or 2.

1 Introduction

A local analytic variety V in a neighborhood of a real point $\xi \in V \cap \mathbb{R}^n$ is said to satisfy the *local Phragmén–Lindelöf condition* $\mathrm{PL}_{\mathrm{loc}}(\xi)$ if and only if there are constants $A > 0$ and $0 < r_1 \leq r_2$ such that every plurisubharmonic function u on $V \cap B(\xi, r_2)$ that satisfies

$$u(z) \leq 1 \quad \text{for } z \in V \cap B(\xi, r_2) \quad \text{and} \quad u(x) \leq 0 \quad \text{for } x \in V \cap \mathbb{R}^n \cap B(\xi, r_2)$$

also satisfies

$$u(z) \le A|\operatorname{Im} z|, \quad z \in V \cap B(\xi, r_1).$$

Here $B(\xi, r)$ denotes the open ball with center at ξ and radius r. When the point ξ is understood, we will call the condition $\operatorname{PL}_{\operatorname{loc}}$.

Classifying the varieties that satisfy $\operatorname{PL}_{\operatorname{loc}}$ is an intrinsically interesting question in complex analysis. Besides this, it is of interest to characterize these varieties because the $\operatorname{PL}_{\operatorname{loc}}$ condition is known to determine the solution to at least two other problems in analysis. In [H], Hörmander showed that a constant coefficient partial differential operator of order m whose principal symbol is the homogeneous polynomial $P_m(z)$ is surjective on the space of all real analytic functions on $\mathbb{R}^n$ if and only if the homogeneous variety $V = V(P_m) := \{z \in \mathbb{C}^n : P_m(z) = 0\}$ satisfies $\operatorname{PL}_{\operatorname{loc}}(\xi)$ at each real point $\xi \in V$ with $|\xi| = 1$. He also gave the first general results on the study of Phragmén–Lindelöf conditions. (Hörmander's results were extended to systems by Andreotti and Nacinovich [AN].) More recently, Vogt [V] has shown that $\operatorname{PL}_{\operatorname{loc}}$ also characterizes the obstruction to the existence of a linear operator that extends real analytic functions from a compact real analytic subvariety $X \subset \mathbb{R}^n$ to all of $\mathbb{R}^n$. He proved that the restriction operator $\rho(f) = f|_X$ of global real analytic functions to X is surjective and has a continuous linear right inverse if and only if the sheaf of germs of analytic functions on X is coherent and X satisfies $\operatorname{PL}_{\operatorname{loc}}(\xi)$ at each of its real points. (The coherence of the sheaf is the condition that characterizes the surjectivity of ρ.)

The aim of this paper is to make explicit a new necessary condition that varieties satisfying $\operatorname{PL}_{\operatorname{loc}}$ must satisfy. The exact definition we give here, in Definition 4.2, is complicated. However, it is one way to make explicit the rough statement, "V is hyperbolic near the real regular points of homogeneous varieties that satisfy the Phragmén–Lindelöf condition." For this reason, we refer to such varieties as *nearly hyperbolic*.

It is clear that varieties satisfying the local Phragmén–Lindelöf condition must have "many real points," since the growth of plurisubharmonic functions is controlled by their upper bounds on the real points. There are two standard ways in which a variety might be said to have many real points. One is in the sense of dimension; i.e., such varieties must have maximal real dimension. In fact they satisfy a stronger property, the strong dimension condition (see Definition 2.2 and Proposition 2.3). However, even this stronger condition is not sufficient. For example, the one-dimensional variety in $\mathbb{C}^2$ given by $x^2 = y^3$ has real dimension 1 at each of its real points but it does not satisfy the Phragmén–Lindelöf condition at the origin ($u(x, y) = |\operatorname{Im} \sqrt{y}|$ is a counter example).

A second, stronger sense of having many real points is when the variety is what one might consider "maximally real"; i.e., hyperbolic near the point. Hörmander showed that if V is *locally hyperbolic at* ξ, then it does satisfy the local Phragmén–Lindelöf condition. Recall that V is locally hyperbolic at ξ if there exists a choice of local coordinates $z = (z'', z')$ with $\{(z'', 0)\}$ transverse to V so that when $z = (\alpha(z'), z') \in V$ is near ξ and z' is real, then $\alpha(z')$ is also real. That is, all the points of V in the fiber of the projection $(z'', z') \to z'$ that lie over real points are real. For varieties of dimension 1, the condition that $\operatorname{PL}_{\operatorname{loc}}$ holds is equivalent to the condition of local hyperbolicity.

However, local hyperbolicity is no longer necessary for varieties of dimension > 1. For example, the two-dimensional varieties in $\mathbb{C}^3$ given by $x^p + y^p + z^p = 0$ with p an odd positive integer satisfy $\mathrm{PL}_{\mathrm{loc}}$ at the origin but are not locally hyperbolic there.

The new condition of being *nearly hyperbolic* that we introduce is intermediate between the dimension condition and local hyperbolicity in strength. We will show here that this new condition is necessary for the Phragmén–Lindelöf condition to hold for varieties of any dimension. Although we will not prove it here, the condition also coincides with the condition of being conically hyperbolic that was introduced in [BMT2]. Consequently, as was shown in that paper, it is also equivalent to the validity of the local Phragmén–Lindelöf estimate for one- or two-dimensional varieties. It seems a reasonable conjecture to us that being nearly hyperbolic is equivalent to the local Phragmén–Lindelöf condition for varieties of all dimensions.

It turns out that there are several different ways to give the definition of a nearly hyperbolic variety. In this paper, we give the one that is most complicated but also the most useful. (It is Definition 4.2 of Section 4.)

In Section 2 we give the precise definition of some local Phragmén–Lindelöf conditions and recall some facts about them that will be used in the proof. In particular, associated to any variety is a large family of varieties obtained by looking at V at small scales. Theorem 2.10 connects $\mathrm{PL}_{\mathrm{loc}}$ for V and for the collection of these families and their limits. In Section 3, some special (transverse or noncharacteristic) coordinate systems are set up to clarify what we mean by saying that "V is near W in a neighborhood $\mathcal{U}$ of a point." The last section contains the precise definition of "the distance from V to W is at most (R, ϵ)" that is used in Definition 4.2 of Section 4. Our main result is Theorem 4.3, which states that varieties satisfying $\mathrm{PL}_{\mathrm{loc}}$ must be nearly hyperbolic.

2 The local Phragmén–Lindelöf condition

In this section, we give the precise definition of the local Phragmén–Lindelöf condition for a variety and discuss how it carries over to a larger family of varieties.

We will always assume that V is an analytic variety of pure dimension k, $1 \leq k < n$, in an open subset $\Omega \subset \mathbb{C}^n$. By a plurisubharmonic (psh) function on V, we mean a function that is locally bounded above on V, that is psh at the regular points of V, and is uppersemicontinuous on V; i.e., $u(z) = \lim \sup_{w \to z} u(w)$. We will also use the notation $B(\xi, r)$ for the Euclidean ball with center ξ and radius r.

Definition 2.1. *The variety V satisfies the local Phragmén–Lindelöf condition* $\mathrm{PL}_{\mathrm{loc}}(\xi)$ *at $\xi \in V \cap \mathbb{R}^n$ if and only if there are constants $0 < r_1 < r_2$ and $A > 0$ such that every function u that satisfies (α) and (β) also satisfies (γ).*

(α) *u is psh and bounded above by 1 on $V \cap B(\xi, r_2)$.*
(β) *$u \leq 0$, $z \in V \cap \mathbb{R}^n \cap B(\xi, r_2)$.*
(γ) *$u(z) \leq A |\operatorname{Im} z|$ for $z \in V \cap B(\xi, r_1)$.*

One can also rephrase Phragmén–Lindelöf conditions in terms of extremal functions defined on $V \cap B(\xi, r_2)$. Set

$$U(z) = U(z; V, B(\xi, r_2)) = \sup\{u(z) : u \text{ satisfies } (\alpha) \text{ and } (\beta)\}. \qquad (2.1)$$

The uppersemicontinuous regularization of U is then psh on V and V satisfies $\mathrm{PL}_{\mathrm{loc}}(\xi)$ if and only if

$$U(z) \le A|\operatorname{Im} z|, \quad z \in V \cap B(\xi, r_1). \qquad (2.2)$$

Let us next recall two properties of the local Phragmén–Lindelöf condition; namely, that varieties satisfying this condition must have many real points in the sense of dimension and that the condition carries over to limits.

Definition 2.2. *A variety V satisfies the* dimension condition *at a real point $\xi \in V$ if and only if there is a neighborhood $\mathcal{U}$ of ξ in which V is the complexification of $V \cap \mathcal{U} \cap \mathbb{R}^n$. It satisfies the* strong dimension condition *at ξ if it satisfies the dimension condition at each real point in a neighborhood of ξ.*

It is equivalent to require that for each irreducible component V_i of V, the real dimension of the real variety $V_i \cap \mathbb{R}^n$ at ξ is equal to the dimension k of the pure dimensional variety V at ξ.

Proposition 2.3. *A variety V that satisfies the local Phragmén–Lindelöf condition at ξ also satisfies the strong dimension condition at that point.*

Proof. See, e.g., [H, Theorem 6.3, p. 177] or [MTV1, Lemma 2.8, p. 519], together with the fact that if V satisfies the local Phragmén–Lindelöf property at ξ, then it also satisfies the property at each point in a neighborhood of ξ. □

By the limit of a sequence (or family) of analytic varieties, we mean in the (equivalent) senses of convergence as currents or as holomorphic k-chains. See Chirka [Ch, especially Chapter 3] for more details or [BMT1], where the convergence is also discussed in more detail.

Proposition 2.4. *Suppose that V_j is a sequence of pure k-dimensional varieties in $B(\xi, r_2)$ that converges to a nonempty limit variety $V_{\lim}$ (which necessarily is of pure dimension k). If each of the varieties V_j satisfies the condition $\mathrm{PL}_{\mathrm{loc}}(\xi)$ with the same constants A and r_1 in Definition 2.1, then so does the limit variety $V_{\lim}$.*

Proof. It is a consequence of [MTV1, Theorem 4.4, p. 524] that the extremal functions defined in (2.1) are lower semicontinuous functions of the variety. That is, as $z_j \in V_j$ converges to a regular point $z \in V$,

$$U(z, V, B(\xi, r_2)) \le \limsup_{j \to \infty} U(z_j, V_j, B(\xi, r_2)).$$

Consequently, $U(z, V, B(\xi, r_2)) \le A|\operatorname{Im} z|$ in $V \cap B(\xi, r_1)$ since the extremal functions for the V_j satisfy this estimate. □

In studying the local Phragmén–Lindelöf principle at a point, it is no loss of generality to suppose the point ξ is the origin, and we will do so in the remainder of the paper. It is clear that the condition is invariant under rigid motions. Therefore, the validity of the Phragmén–Lindelöf condition immediately induces a large family of varieties that also satisfy the condition.

Definition 2.5. *Denote by $\mathcal{T}$ the collection of all affine transformations of $\mathbb{C}^n$ of the form*

$$T(w) = T_{a,r,U}(w) = a + rUw, \quad a \in \mathbb{R}^n, \quad 0 < r \leq |a|, \tag{2.3}$$

where U is a real, orthogonal linear transformation. Also, denote by $\mathcal{T}_\epsilon$ the collection of all such transformations where $|a| \leq \epsilon$. Similarly, we set

$$V_T = T^* V := \{w : T(w) \in V\}$$

and

$$\mathcal{V}_\epsilon = \{V_T : T \in \mathcal{T}_\epsilon\}. \tag{2.4}$$

Since $0 \in V \subset \Omega$, the variety $T^* V = T^*_{a,r,U} V$, the pullback of V under such an affine transformation, is defined on a large subset of $\mathbb{C}^n$ that contains $|w| \leq R(\epsilon) \to \infty$ when $T \in \mathcal{T}_\epsilon$ with $\epsilon \to 0$. So, if we take the limit of a convergent family of the varieties V_T as $a \to 0$, then the limit variety will be defined in all of $\mathbb{C}^n$ (remember that $r \leq |a|$). The variety V_T is obtained by looking at the little piece of V that lies in $B(a, r)$ and then rescaling it to the unit ball. For this reason, we will call a limit of these varieties a *blowup* of V.

Definition 2.6. *Define $\mathcal{B}(V) = \mathcal{B}(V, 0)$ to be the set of all varieties $V_{\lim}$ such that there exists $T_j \in \mathcal{T}_{\epsilon_j}$ with $\epsilon_j \to 0$ and*

$$V_{\lim} = \lim_{j \to \infty} T_j^* V.$$

Remark. Although we will not use it here, these limit varieties are always algebraic. A proof can be given by the same argument used to prove [BMT1, Corollary 3.5, p. 69].

We next observe that all the varieties in these families, including the blowups, satisfy the Phragmén–Lindelöf principle in a neighborhood of every real point, with a uniform choice of the constant A.

Proposition 2.7. *The variety V satisfies $\mathrm{PL}_{\mathrm{loc}}$ if and only if for each $\xi \in \mathbb{R}^n$ and each $R > 0$, there exist $\epsilon_0 > 0$ and $A_1 > 0$ such that the following holds: For each variety V_T in $\mathcal{V}_{\epsilon_0}$, each function $v_T : V_T \to [-\infty, +\infty)$ that satisfies (α) and (β) also satisfies (γ):*

(α) v_T *is psh and bounded above by 1 on $V_T \cap B(\xi, R)$.*

(β) $v_T(z) \leq 0$, $z \in V_T \cap \mathbb{R}^n \cap B(\xi, R)$.

(γ) $v_T(w) \leq A_1 |\operatorname{Im} w|$ for $w \in V_T \cap B(\xi, \frac{R}{2})$.

In fact, ϵ_0 can be taken to be any number smaller than $r_1/(1 + |\xi| + R)$ and $A_1 = 4r_2A/R$, where $0 < r_1 < r_2$ and A are the constants in Definition 2.1.

Proof. The proof that the families $\mathcal{V}_\epsilon$ satisfy the Phragmén–Lindelöf estimate with a uniform constant follows from a localization argument due to Hörmander. The choice of ϵ_0 described in the proposition guarantees that each $T \in \mathcal{T}_\epsilon$ maps the balls $w \in B(\xi, R)$ into the ball $|z| < r_1$. Therefore, each psh function v_T on $V_T \cap B(\xi, R)$ determines a psh function φ on $V \cap T(B(\xi, R))$, the part of V inside a small ball in $|z| \leq r_1$ by the formula

$$\varphi(z) = \varphi(a + rUw) = r v_T(w).$$

The function φ "extends" to a psh function on all of V by the following localization argument.

Let $H(z)$ denote a psh function on the unit ball $|z| \leq 1$ in $\mathbb{C}^n$ such that $H(z) \leq |\operatorname{Im} z|$, $H(iy) \geq 0$, and $H(z) \leq |\operatorname{Im} z| - \frac{1}{2}$ for $|z| = 1$. For example, $H(z) = \frac{1}{2}(|\operatorname{Im} z|^2 - |\operatorname{Re} z|^2)$. Pick a fixed point $w_0 \in V_T \cap B(\xi, R/2)$ where we want to prove the estimate of (γ) under the hypotheses that v_T satisfies (α) and (β). Set $c = \operatorname{Re}(T(w_0))$ and then let

$$\widetilde{\varphi}(z) = \varphi(z) + 2r H\left(\frac{2(z - c)}{rR}\right).$$

The psh function $\widetilde{\varphi}$ satisfies $v_T(w_0) \leq \widetilde{\varphi}(T(w_0))$ since $H(iy) \geq 0$. From the estimate for H on the boundary of the unit ball, the function $\widetilde{\varphi}$ satisfies

$$\widetilde{\varphi}(z) \leq r + 2r\left(\frac{2|\operatorname{Im} z|}{rR} - \frac{1}{2}\right) \leq \frac{4|\operatorname{Im} z|}{R} \quad \text{for } |z - c| = rR/2.$$

Therefore, if we replace $\widetilde{\varphi}$ by the maximum of itself and $4|\operatorname{Im} z|/R$, then the resulting psh function extends to be psh on all of V by defining it to be equal to $4|\operatorname{Im} z|/R$ outside this ball. In particular, it is bounded by $4r_2/R$ on the ball of radius r_2. And since v_T is ≤ 0 at all the real points of V_T, this extended function vanishes on all the real points of V. From the local Phragmén–Lindelöf condition applied to $R/4r_2$ times this function, we conclude that the extended function is bounded by $4r_2A|\operatorname{Im} z|R$ on the ball $|z| \leq r_1$. Applying this at the point $z_0 = T(w_0)$ then gives the inequality

$$v_T(w_0) \leq \frac{4r_2A|\operatorname{Im} w_0|}{R},$$

which is what was to be shown.

We omit the straightforward proof of the converse part of the proposition. $\quad\square$

Corollary 2.8. *The conclusions of Proposition 2.7 also apply with the varieties V_T replaced by any variety in $\mathcal{B}(V)$.*

Proof. This is a direct consequence of the proposition, the definition of the varieties in $\mathcal{B}(V)$ as limits of those in $\mathcal{V}_\epsilon$, and Proposition 2.4.   □

Corollary 2.9. *It is equivalent in Proposition* 2.7 *to assume that the result holds only for all the points* $\xi \in \mathbb{R}^n$ *that lie in some limit variety of* $\mathcal{B}(V)$.

If a real point ξ is not a point in some one of the limit varieties in $\mathcal{B}(V)$, then it has a neighborhood that is disjoint from the families $\mathcal{T}(V)_\epsilon$ for sufficiently small $\epsilon > 0$. In this case, all of the conditions of Proposition 2.7 are vacuous.

Remark. Obtaining this family is the reason for the restriction $r \leq |a|$ in the definition of the linear transformations, T. If r is large with respect to $|a|$, then the only limit variety obtained is $T_0(V)$, the tangent variety to V at 0. If $r \sim |a|$, then one obtains translates of $T_0(V)$, but if $r << |a|$, many new varieties can arise.

The combination of Proposition 2.7 and Corollary 2.9 gives us the following theorem, which shows there are (too) many conditions that must be satisfied when the Phragmén–Lindelöf condition holds. One hopes to study the problem by identifying which of these are crucial for the validity of $\mathrm{PL}_{\mathrm{loc}}$.

Theorem 2.10. *The variety* $0 \in V$ *of pure dimension* $1 \leq k \leq n$ *satisfies* $\mathrm{PL}_{\mathrm{loc}}$ *at the origin if and only if for each* $V_{\lim} \in \mathcal{B}(V)$ *and each real point* $\xi \in V_{\lim}$, *there exist* $0 < R_1 < R_2$, $\epsilon_0 > 0$, *and a constant* $A_1 > 0$ *such that the following holds: For each variety* V_T *in* $\mathcal{V}_{\epsilon_0}$, *each function* $v_T : V_T \to [-\infty, +\infty)$ *that satisfies* (α) *and* (β) *also satisfies* (γ):

(α) v_T *is psh and bounded above by 1 on* $V_T \cap B(\xi, R_2)$.
(β) $v_T(z) \leq 0$, $z \in V_T \cap \mathbb{R}^n \cap B(\xi, R_2)$.
(γ) $v_T(w) \leq A_1 |\operatorname{Im} w|$ *for* $w \in V_T \cap B(\xi, R_1)$.

3 Transverse projections, homogeneous varieties, and the EH property

To describe hyperbolicity conditions for varieties we need to make use of *transverse* or *noncharacteristic* coordinate systems. We keep the assumption throughout this section that V is a variety of pure dimension k in a neighborhood of the origin in $\mathbb{C}^n$. For details of the results we use about varieties, the excellent book of Chirka [Ch] is a good reference.

Recall that a (real, linear) projection map $\pi : \mathbb{C}^n \to \mathbb{C}^n$ is said to be *transverse* to V if the following hold:

(1) π has rank k, and the kernel and the image of π are spanned by real vectors.
(2) The kernel of π intersects the cone $T_0(V)$ of tangents to V at the origin only at the point $\{0\}$.

One can then choose coordinates $z = (z'', z') \in \mathbb{C}^{n-k} \times \mathbb{C}^k$ such that $\pi(z'', z') = z'$ and in these coordinates V can be described in a polydisk $\Delta = \Delta'' \times \Delta' = \{|z''| < \delta'', \;\; |z'| < \delta'\}$ as a branched analytic covering,

$$V = \{(\alpha_j(z'), z') : 1 \leq j \leq m, \; |z'| < \delta'\}, \tag{3.1}$$

where m is the multiplicity of V at 0. The transversality condition means that there is a constant $K = K(V, \pi)$ such that

$$|\alpha_j(z')| \leq K|z'|, \quad |z'| < \delta'.$$

There is a *branch locus* $B'_\pi(V) \subset \Delta'$ where the points in the fiber over z' are not all distinct, and a corresponding set of branch points $B_\pi(V)$ in V, the points where $(\alpha_j(z'), z') = (\alpha_l(z'), z')$ for some $j \neq l$. Thus we have

$$\alpha_j(z') \neq \alpha_l(z') \quad \text{if } 1 \leq j \neq l \leq m, \quad z' \in \Delta' \setminus B'_\pi(V).$$

Next suppose that W is a *homogeneous* variety of pure dimension k; i.e., $\zeta z \in W$ if $z \in W$ and $\zeta \in \mathbb{C}$. A particularly important example for us will be $W = T_0(V)$, or W might be the tangent cone to one of the limit varieties $V_{\lim} \in \mathcal{B}(V)$ at a real point $\xi \in V_{\lim}$. Note that in such examples, if the limit variety is taken in the sense of currents or holomorphic chains, then the limit variety will have the same multiplicity as the original variety. For example, if

$$T_0[V] = \lim_{t \to 0} \frac{1}{t}[V],$$

where $[V]$ denotes the positive current of bidimension (k, k) associated to V, then as a holomorphic chain, $T_0[V]$ will also have the same multiplicity m as V at 0. However, this limit may have repeated irreducible components so that the homogeneous variety W that is its support may have a smaller multiplicity. That is,

$$W = \{(\gamma_i(z'), z') : 1 \leq i \leq \sigma\}, \quad z' \in \mathbb{C}^k,$$

where σ is the multiplicity of W at 0, or what is the same thing, the *degree* of W as a global algebraic variety. The hypothesis that the projection is transverse to W means that there is a constant K, depending on W and π such that

$$|\gamma_i(z')| \leq K|z'|, \quad z' \in \mathbb{C}^k. \tag{3.2}$$

In case $W = T_0(V)$, the $\gamma_i(z')$ are limit points of $\alpha_j(\zeta z')/\zeta$ as $\zeta \to 0$.

Associated to a homogeneous variety W, a transverse projection π, and a positive number $\eta > 0$, we consider the cones of points in $\mathbb{C}^k$ and W where all the points in the fiber of π are distinct and separated by at least $\eta|z'|$. That is,

$$\Gamma'(\eta) := \Gamma'_\pi(\eta, W) := \{z' : |\gamma_i(z') - \gamma_j(z')| > \eta|z'|, \; 1 \leq i \neq j \leq \sigma\}, \tag{3.3}$$

$$\Gamma(\eta) := \Gamma_\pi(\eta, W) := W \cap \pi^{-1}(\Gamma'_\pi(\eta)). \tag{3.4}$$

Note that, given a simply connected open subset of $\Gamma'(\eta)$, the functions $\gamma_i(z')$ can be defined as σ individual single-valued analytic functions on this set.

Definition 3.1. *A homogeneous variety W is said to be of class EH if and only if for each transverse projection π and each $\eta > 0$, there is a constant $c > 0$ such that for each $x' \in \Gamma'_\pi(\eta) \cap \mathbb{R}^k$ and each $1 \le j \le \sigma$, either*

$$| \operatorname{Im} \gamma_j(x')| \ge c|x'| \quad or \quad \operatorname{Im} \gamma_j(x') = 0.$$

That is, the points in W either look "elliptic" or "hyperbolic." It is pretty easy to recognize when a homogeneous variety has the EH property; it is very close to the dimension condition.

Proposition 3.2. *A homogeneous variety W is of class EH if and only if it satisfies the dimension condition at each real regular point $0 \ne z \in W$. In particular, every homogeneous variety that satisfies $\mathrm{PL_{loc}}(0)$ is of class EH.*

Proof. First, suppose that W is of class EH. If $z_0 \ne 0$ is a real, regular point of W, then there exist a transverse projection and an $\eta > 0$ such that $z_0 = (\gamma_i(x'_0), x'_0)$ where $x'_0 \in \Gamma'(\eta) \cap \mathbb{R}^k$. There is a unique, single-valued analytic function $\gamma_i(x')$ that determines the branch of W that passes through z_0. For x' real and near x'_0, then from the EH condition, $\gamma_i(x')$ must also be real since $\gamma_i(x'_0)$ is real. So, W must also have real dimension k at z_0 as asserted by the dimension condition.

To see the converse, let $S'(\eta) = \{x' \in \mathbb{R}^k \cap \Gamma'(\eta) : |x'| = 1, \operatorname{Im} \gamma_i(x') \ne 0$ for some $\gamma_i(x')\}$ and $c(\eta) = \min\{|\operatorname{Im} \gamma_i(x')| : x' \in S'(\eta), \operatorname{Im} \gamma_i(x') \ne 0\}$. Because W is homogeneous, the assertion that W is of class EH is simply that $c(\eta) > 0$ (or $S'(\eta)$ is empty) for each $\eta > 0$. If this were false, there would have to exist points $(x'_j, \gamma_i(x'_j))$ with $x'_j \in S'(\eta)$ that converge to a real point $(x'_\infty, \gamma_i(x'_\infty))$ with $\operatorname{Im} \gamma_i(x'_\infty) = 0$. Then x'_∞ is in the closure of $S'(\eta) \subset\subset \Gamma'(\eta/2)$, so there is a unique branch of W through this limit point. Since W satisfies the dimension condition at this real, regular point, this unique branch must have $\gamma_i(x')$ real for real x' near the limit point, contrary to the fact that the limit point was to be a limit of points with $\operatorname{Im} \gamma_i(x_j) \ne 0$. Thus the EH property must hold.

The last assertion of the proposition is a consequence of Proposition 2.3.   □

4 A measure of distance between varieties

Suppose V and W are two varieties of pure dimension k that are "close." A basic question of complex analysis is to determine how closely related are basic analytic objects, such as plurisubharmonic functions, on V and W. If there is an analytic map of V to W then this map can be used to pull back functions from W to V. It can be used to push forward functions from V to W while preserving bounds for the functions. Further, if all the preimages of real points are real, i.e., the map is hyperbolic, then it will push forward functions that vanish on real points to functions that again vanish on real points. We would like to have such a map, especially for subfamilies of the varieties V_T that converge to $V_{\lim} \in \mathcal{B}(V)$.

It is easy to see that such maps can be locally defined near regular points of the limit variety; one can take the coordinate projection along any subspace transverse to the limit. However, it does not seem to be the case that such a map can always exist when the limit variety has singularities. The following definition is to make explicit the sets on which one can "line up" points of V_T with points of a limit variety $V_{\lim}$ and to estimate the (small) exceptional set where this is not possible. While we do not do so here, one can explain how to "calculate" from the varieties V and W the measures R, ϵ given in the definition.

Recall that $B(x, R)$ denotes the ball with center at x and radius R.

Definition 4.1. *Let V, W be analytic varieties of pure dimension $k \geq 1$ in an open set $\Omega \subset \mathbb{C}^n$. Suppose $\xi \in W$ and let $\mathcal{U} \subset \Omega$ be a bounded neighborhood of ξ. Then we say that V is of distance at most (R, ϵ) from W in $\mathcal{U}$ if and only if*

$$V \cap \mathcal{U} \subset B(\xi, R) \cup \bigcup_{z \in W \cap \mathcal{U}} B(z, \epsilon|z - \xi|). \tag{4.1}$$

We will denote the smallest such R with this property by

$$d(V; W, \epsilon, \mathcal{U}) = \inf\{R : (4.1) \text{ is satisfied}\}.$$

It may happen that $R = d(V; W, \epsilon, \mathcal{U}) = 0$. This is the case if and only if V is contained in a small "cone" around W; i.e.,

$$V \cap \mathcal{U} \subset \bigcup_{z \in W \cap \mathcal{U}} B(z, \epsilon|z - \xi|). \tag{4.2}$$

The minimum value of R is a measure of the amount by which this inclusion fails.

This definition is not symmetric with respect to V and W. It does imply that V is near a subvariety of W when ϵ and R are small.

We will be studying the condition near a point $\xi \in W$ which we will henceforth assume is $\xi = 0 \in W$. While the definition as given is independent of the choice of coordinates, it is much more useful for our purposes to have it in terms of the coordinates associated to a real, linear projection map $\pi : \mathbb{C}^n \to \mathbb{C}^n$ that is transverse to W at 0. Let $\pi : \mathbb{C}^n \to \mathbb{C}^n$ be such a projection. We will continue to use the coordinates introduced in Section 3. Let $\delta'' > 0$, $\delta' > 0$ be numbers such that the polydisk

$$\Delta(\delta'', \delta') := \Delta''(\delta'') \times \Delta'(\delta') := \{(z'', z') : |z''| < \delta'', \ |z'| < \delta'\}$$

about $0 \in W$ is a relatively compact subset of Ω, the open set in which W is an analytic variety. We will also suppose that the polydisk is chosen small enough that $\pi : W \cap \Delta(\delta'', \delta') \to \Delta(\delta')$ is a proper branched analytic covering map over $\Delta'(\delta')$. If the multiplicity of W at the origin is σ, then we can write

$$W = \{(\gamma_i(z'), z') : 1 \leq i \leq \sigma\},$$

where the $\gamma_i(z')$ are distinct for $z' \in \Delta'(\delta')$, except over the proper subvariety $B'_\pi(W)$ of $\Delta'(\delta')$. Further, there are constants $K > 0$, $\epsilon_0 > 0$ such that

$$|\gamma_i(z')| \le K|z'|, \qquad z' \in \Delta'(\delta'), \quad 1 \le i \le \sigma, \tag{4.3}$$

$$|\gamma_i(z')| \le (\delta'' - \epsilon_0)\frac{|z'|}{\delta'}, \quad z' \in \Delta'(\delta'), \quad 1 \le i \le \sigma.$$

Therefore, if $\Delta := \Delta(\delta'', \delta') \subset \mathcal{U}$, if $d(V; W, \epsilon, \mathcal{U}) < \delta''$, and if $\epsilon < \epsilon_0$, then $\pi : V \to \Delta(\delta')$ is also a proper branched covering map. Consequently, if μ is the degree of this map over $\Delta(\delta')$, we can also represent V in the form

$$V = \{(\beta_i(z'), z') : 1 \le i \le \mu\},$$

where the $\beta_i(z')$ are distinct except for z' in $B'_\pi(V)$.

With this notation, we next define a nearly hyperbolic family of varieties, and by this, a nearly hyperbolic variety.

Definition 4.2. *Let $\{V_T\}_{T \in \mathcal{T}}$ be a family of varieties in a neighborhood of a real point $\xi \in \mathbb{C}^n$, which, without loss of generality, we can take to be the origin. We say that the family is* nearly hyperbolic *at ξ if the following condition holds for every homogenous variety $W \subset \mathbb{C}^n$ that has pure dimension k, is of class EH, and satisfies the dimension condition at 0: For every real linear projection π that is transverse to W, and for every $\eta > 0$, there exists a choice of constants $\epsilon_0 > 0$, $R_0 > 0$, $C > 0$, and $\delta'' > 0$, $\delta' > 0$, depending only on W, η, π, such that, for every choice of $0 < \epsilon < \epsilon_0$ and $0 < R < R_0$, condition (3) holds whenever conditions (1) and (2) hold.*

(1) *$V_T \in \{V_T\}_{T \in \mathcal{T}}$ is of distance at most (R, ϵ) from W in $\Delta(\delta'', \delta')$.*
(2) *$z = (\beta, x') \in V_T$ satisfies $x' = \pi(z) \in \mathbb{R}^k \cap \Gamma'_\pi(\eta)$ and $CR \le |x'| < \delta'/4$.*
(3) *There exists a unique $w = (\gamma_i, x') \in W$ in the fiber of W over x' that is closer to $z = (\beta, x')$ than any other point in the fiber and*

$$\operatorname{Im} \beta = 0 \quad if \operatorname{Im} \gamma_i = 0.$$

Further, we say that the variety V is nearly hyperbolic at 0 if for each real point ξ in a neighborhood of the origin, there is $\epsilon > 0$ such that the family V_ϵ is nearly hyperbolic at ξ.

The connection with the study of the local Phragmén–Lindelöf condition is then provided by the following two theorems.

Theorem 4.3. *If V satisfies $\mathrm{PL}_{\mathrm{loc}}$ at the origin, then V is nearly hyperbolic at the origin.*

Theorem 4.4. *If the family $\{V_T\}_{T \in \mathcal{T}}$ is nearly hyperbolic and has a unique limit variety V_{lim} that satisfies the condition $\mathrm{PL}_{\mathrm{loc}}(\xi)$ at each of its real points, then for each $\xi \in V_{\mathrm{lim}} \cap \mathbb{R}^n$, there are constants $\epsilon_0 > 0$, $R_0 > 0$, $A > 0$ and $B > 0$ and $0 < r_1 < r_2$ such that (3) holds whenever (1) and (2) are satisfied.*

(1) *$0 < \epsilon \le \epsilon_0$, $0 \le R \le R_0$, and V_T is of distance at most (R, ϵ) from W, the tangent cone to V_{lim} at ξ.*

(2) v_T is psh on $V_T \cap B(\xi, r_2)$, with $v_T \leq 1$ and $v_T \leq 0$ on $V_T \cap B(\xi, r_2) \cap \mathbb{R}^n$.
(3) $v_T(z) \leq A|z - \xi| + BR$ for $z \in V_T \cap B(\xi, r_1)$.

Theorem 4.3 shows that the nearly hyperbolic condition is necessary, while Theorem 4.4 shows that some Phragmén–Lindelöf-type bounds hold, modulo certain errors. It is really a microlocal version of the Phragmén–Lindelöf estimates. In many (perhaps all?) cases, it is possible to apply the estimates of Theorem 4.4 (many times) to deduce that the converse holds; i.e., V satisfies $\mathrm{PL_{loc}}$ at the origin. This is the method that has been successfully applied to prove the equivalence of the two conditions for surfaces.

We will give the proof of Theorem 4.3 below. Theorem 4.4 is not yet published, and there is not space available here to give the proof.

For the proof of Theorem 4.3, we need a restatement of the condition on the distance from V to W in terms of the fibers of a transverse projection.

Proposition 4.5. *Let ϵ_0 and K be as in (4.3). If*

(i) $0 < \epsilon < \min\{\epsilon_0, 1/2(1 + K)\}$,
(ii) $R = d(V; W, \epsilon, \Delta(\delta'', \delta'))$,

then there is a constant $C > 0$ that depends only on the dimension n and the multiplicity σ of W at the origin such that for all z' with $|z'| \leq \frac{1}{4}\delta'$,

$$\max_{1 \leq j \leq \mu} \min_{1 \leq i \leq \sigma} |\beta_j(z') - \gamma_i(z')| \leq \max\{(1 + K)R, C\epsilon^{1/\sigma}|z'|\}.$$

In the converse direction, if

$$\max_{1 \leq j \leq \mu} \min_{1 \leq i \leq \sigma} |\beta_j(z') - \gamma_i(z')| \leq R + \epsilon|z'|,$$

then for every $\lambda \geq 1$,

$$d(V; W, (1 + \lambda)\epsilon, \Delta(\delta'', \delta')) \leq \left(1 + \frac{1}{\lambda} + \frac{1 + K}{\lambda\epsilon}\right)R.$$

Proof. The converse direction is easy to see so we prove it first. If $\lambda\epsilon|z'| \geq R$, then the hypothesis implies that for some i,

$$|(\beta_j(z'), z') - (\gamma_i(z'), z')| = |\beta_j(z') - \gamma_i(z')| \leq R + \epsilon|z'| \leq (1 + \lambda)\epsilon|z'|.$$

On the other hand, if $\lambda\epsilon|z'| \leq R$, then

$$|(\beta_j(z'), z')| \leq |\beta_j(z') - \gamma_i(z')| + |(\gamma_i(z'), z')|$$

$$\leq R + \epsilon|z'| + (1 + K)|z'| \leq \left(1 + \frac{1}{\lambda} + \frac{1 + K}{\lambda\epsilon}\right)R,$$

which is the meaning of the last inequality of the proposition.

In the other direction, if $(\beta_j(z'), z')$ lies in the closed ball of radius R about the origin, then for any i, $|\beta_j(z') - \gamma_i(z')| \leq R + |\gamma_i(z')| \leq R + K|z'| \leq (1 + K)R$.

Otherwise, from the property that V is of distance at most (R, ϵ) from W, there exists a point w' and a choice of i such that

$$|(\gamma_i(w'), w') - (\beta_j(z'), z')| \le \epsilon |(\gamma_i(w'), w')| \le \epsilon(1 + K)|w'|. \tag{4.4}$$

We have to show that the point $(\gamma_i(w'), w') \in W$ can be changed to one with w' replaced by z' while keeping almost the same estimate.

The last inequality implies that $|z' - w'| \le \epsilon(1 + K)|w'|$ so that $(1 - \epsilon(1 + K))|w'| \le |z'|$ and therefore

$$|\gamma_i(w') - \beta_j(z')| \le \frac{\epsilon(1 + K)}{1 - \epsilon(1 + K)} |z'|.$$

Therefore, the estimate of the first part of the proposition will be proved with

$$C = \frac{(1 + K)}{1 - \epsilon(1 + K))} + C_1 \le 2(1 + K) + C_1$$

provided we can show that there is an index l, $1 \le l \le \sigma$, and a constant $C_1 > 0$ such that

$$|\gamma_i(w') - \gamma_l(z')| \le C_1 \epsilon^{1/\sigma} |z'|. \tag{4.5}$$

To see this, consider the canonical defining function of W,

$$P(z, \zeta) := P(z'', z', \zeta, W) := \prod_{l=1}^{\sigma} \langle z'' - \gamma_l(z'), \zeta \rangle, \tag{4.6}$$

where $\zeta \in \mathbb{C}^{n-k}$. Because of the upper bound of (4.3), P has the upper bound

$$|P(z, \zeta)| \le (|z''| + K|z'|)^{\sigma} |\zeta|^{\sigma}, \quad |z'| \le \delta', \zeta, z'' \in \mathbb{C}^{n-k},$$

so that by Cauchy's inequality for derivatives, the gradient of P with respect to the z' variables satisfies for every ρ with $2\rho < \delta'$,

$$|\nabla_{z'} P(z, \zeta)| \le \frac{(|z''| + 2K\rho)^{\sigma} |\zeta|^{\sigma}}{\rho}, \quad |z'| \le \rho.$$

Therefore, again using (4.3), we have, whenever both $|z'|$ and $|w'|$ are smaller than ρ,

$$|P(\gamma_i(w'), z', \zeta)| = |P(\gamma_i(w'), z', \zeta) - P(\gamma_i(w'), w', \zeta)| \\ \le (3K|\zeta|)^{\sigma} \rho^{\sigma-1} |z' - w'| \le 2\epsilon(1 + K)(3K|\zeta|\rho)^{\sigma}, \tag{4.7}$$

where the last inequality results from the estimate for $|z' - w'|$ in (4.4) and $|w'| \le \rho$. On the other hand, we can get a lower bound for $|P(\gamma_i(w'), z', \zeta)|$ for certain values of ζ in terms of the minimum distance from $\gamma_i(w')$ to one of the $\gamma_l(z')$ as follows. For every vector $u \in \mathbb{C}^{n-k}$, the set of vectors ζ on the unit sphere such that $|\langle u, \zeta \rangle| < \eta|u|$ is a set of measure at most $C_2\eta$ for some constant $C_2 > 0$. Therefore, the set of ζ with $|\langle \gamma_i(w') - \gamma_l(z'), \zeta \rangle| \le \eta|\gamma_i(w') - \gamma_l(z')|$ for some $1 \le l \le \sigma$ has measure at most

$\sigma C_2 \eta$. Consequently, if η is small enough that $\sigma C_2 \eta$ is less than the surface area of the unit sphere in $\mathbb{C}^{n-k}$, that is,

$$0 < \eta < \frac{2\pi^n}{\sigma C_2 (n-1)!}, \tag{4.8}$$

then we can find a vector ζ_0 such that none of the last inequalities hold. That is, for each choice of z', w', a vector $\zeta_0 \in \mathbb{C}^{n-k}$ of modulus 1 exists such that

$$|P(\gamma_i(w'), z', \zeta_0)| \geq \eta^\sigma \prod_{l=1}^\sigma |\gamma_i(w') - \gamma_l(z')| \geq \eta^\sigma \left[\min_{1 \leq l \leq \sigma} |\gamma_i(w') - \gamma_l(z')| \right]^\sigma.$$

Combining this inequality with that of (4.7) and taking the σth root of both sides then gives

$$\min_{1 \leq l \leq \sigma} |\gamma_i(w') - \gamma_l(z')| \leq \epsilon^{1/\sigma}(1+K)^{1/\sigma}\frac{3K}{\eta}\rho \quad \text{if } |z'|, |w'| \leq \rho < \frac{1}{2}\delta'.$$

It was observed earlier that $1 - \epsilon(1 + K))|w'| \leq |z'|$. Consequently, if we choose $\rho = |z'|/(1 - \epsilon(1 + K)) < \frac{1}{2}\delta'$, then the inequality holds. Therefore, (4.5) is valid with the constant

$$C_1 = \frac{3K(1+K)^{1/\sigma}}{\eta(1 - \epsilon(1+K))}$$

whenever $|z'| < (1 - \epsilon(1 + K))\frac{1}{2}\delta' \leq \frac{1}{4}\delta'$. This completes the proof. $\qquad\square$

Proof of Theorem 4.3. First, observe that if V satisfies $\mathrm{PL}_{\mathrm{loc}}$ at the origin, then the families $\mathcal{V}_\epsilon$ all satisfy the Phragmén–Lindelöf condition with a uniform constant A. Thus, in proving Theorem 4.3, it is enough to deal with a single but arbitrary variety $V = V_T$ of this family, one that satisfies the distance condition of (1) of Definition 4.2 with respect to the homogeneous variety W given there. We will assume in the remainder of the proof that V is such a variety and that (1) and (2) are satisfied.

Begin by letting ϵ_0 be equal to the minimum of the constant from Proposition 4.5 and $\eta/16$. The number C also comes from Proposition 4.5. Fix a point satisfying the conditions in (2), say, $z_0 = (\beta_0, x_0')$. Then from Proposition 4.5, the distance from β_0 to one of the points $\gamma_i(x_0')$ is at most $C\epsilon^{1/\sigma}|x_0'|$. Decreasing the choice of ϵ_0 if necessary, we can assume this number is smaller than $\eta|x_0'|/16$. Since the points in the fiber of W over x_0' are at least distance $\eta|x_0'|$ apart, the uniqueness of $\gamma_i(x_0')$ nearest to β_0 is clear. Further, there is a radius $r = r(\eta) > 0$ depending only on η and W such that the ball $|z' - x_0'| < r|x_0'|$ is contained in $\Gamma'(\eta/2)$. Over this ball, all the branches γ_j can be chosen as single-valued analytic functions. Further, over this disk, the map that sends $(\beta, z') \in V_T$ to the closest point of W in the fiber over z' is a single-valued analytic map, the " nearest point in the fiber map." Denote this map by N_π.

The assertion of the theorem is that z is real if $N_\pi(z)$ is real. To prove this, consider the function defined on the branches of V_T over the ball $|z' - x_0'| < r$ nearest to the branch $(\gamma_i(z'), z')$ of W by

$$\varphi(z) = (A+1)\varphi(\beta, z') = (A+1)|\operatorname{Im}(\beta - \gamma_i(z'))| = (A+1)|\operatorname{Im}(z - N_\pi(z))|.$$

The function φ is bounded on this set by $(A+1)C\epsilon|z'|$ which we can assume is bounded by $1/2$, possibly by shrinking the choice of ϵ_0. Therefore, if we argue as in the proof of Proposition 2.7 and replace φ by

$$\widetilde{\varphi} = \max\left\{\varphi(z) + rH\left(\frac{z' - x_0'}{r}\right), |\operatorname{Im} z|\right\},$$

then we obtain a psh function that is no greater than $|\operatorname{Im} z|$ on these branches of V_T. Therefore, the function can be extended to be psh on all of V_T by setting it equal to $|\operatorname{Im} z|$ at all other points of V_T. The resulting function is bounded by 1 and vanishes at the real points of V_T, so by the hypothesis that V_T satisfies the Phragmén–Lindelöf condition with constant A, we obtain the bound

$$(A+1)|\operatorname{Im}\beta_0| = \varphi(\beta_0, x_0') \le A|\operatorname{Im}(\beta_0, x_0')| = A|\operatorname{Im}\beta|,$$

which can hold only if $\operatorname{Im}\beta_0 = 0$, as asserted.   □

References

[AN]   A. Andreotti and M. Nacinovich, *Analytic Convexity and the Principle of Phragmén–Lindelöf*, Pubblicazioni della Classe di Scienze: Quaderni, Scuola Normale Superiore Pisa, Pisa, 1980.

[BMT1]   R. W. Braun, R. Meise, and B. A. Taylor, Higher order tangents to analytic varieties along curves, *Canad. J. Math.*, **55** (2003), 64–90.

[BMT2]   R. W. Braun, R. Meise, and B. A. Taylor, The geometry of analytic varieties satisfying the local Phragmén–Lindelöf condition and a geometric characterization of the partial differential operators that are surjective on $A^\omega(\mathbf{R}^4)$, *Trans. Amer. Math. Soc.*, **356** (2003), 1315–1383.

[Ch]   E. M. Chirka, *Complex Analytic Sets*, Kluwer, Dordrecht, the Netherlands, 1989.

[H]   L. Hörmander, On the existence of real analytic solutions of partial differential equations with constant coefficients, *Invent. Math.*, **21** (1973), 151–183.

[MTV1]   R. Meise, B. A. Taylor, D. Vogt, Extremal plurisubharmonic functions of linear growth on algebraic varieties, *Math. Z.*, **219**-4 (1995), 515–537.

[MTV2]   R. Meise, B. A. Taylor, D. Vogt, Phragmén–Lindelöf principles on algebraic varieties, *J. Amer. Math. Soc.*, **11** (1998), 1–39.

[V]   D. Vogt, Continuous linear extension operators for real analytic functions on varieties, preprint.

Sampling and Local Deconvolution

David F. Walnut

Department of Mathematical Sciences
George Mason University
Fairfax, VA 22030
USA
dwalnut@gmu.edu

Summary. In a series of papers starting in the late 1970s, Carlos Berenstein and collaborators explored various aspects of the problem of recovering a function locally from its convolutions with compactly supported distributions. These problems have variously been referred to as *multisensor deconvolution problems*, *analytic and polynomial Bezout problems*, and *local Pompeiu problems*. The focus of Carlos' work in this area has been to find explicit and computable solutions. The purpose of this paper is to present a point of view on a small subclass of these problems in which some classical techniques from the theory of sampling and interpolation in Paley–Wiener spaces of entire functions can shed some light on their explicit solution.

1 Introduction

In a series of papers beginning in the late 1970s [2, 3, 4, 5, 6, 7, 8, 9, 10, 11, 12, 17] Carlos Berenstein and collaborators (most notably A. Yger, B. A. Taylor, and R. Gay) investigated, among many other things, the following two problems:

(A) *The multisensor deconvolution problem.* Given a collection of compactly supported distributions $\{\mu_k\}_{k=1}^m \subseteq \mathcal{E}'(\mathbf{R}^d)$, find explicit formulas for compactly supported distributions $\{\nu_k\}_{k=1}^m \subseteq \mathcal{E}'(\mathbf{R}^d)$ such that

$$\sum_{k=1}^m \mu_k * \nu_k = \delta. \tag{1}$$

(B) *The local Pompeiu problem* (see [45] and [46]). Given a collection of compact sets $\{E_k\}_{k=1}^m \subseteq \mathbf{R}^d$, symmetric about the origin, let $E = \sum_{k=1}^m E_k$, let U be an open set containing E and define the operator[1] $P: C(U) \to \bigoplus_{k=1}^m C(U - E_k)$ by

$$Pf = \left(f * \chi_{E_1} \big|_{U-E_1}, \dots, f * \chi_{E_m} \big|_{U-E_m} \right). \tag{2}$$

---

[1] For an open set $\mathcal{O}$, $C(\mathcal{O})$ is the space of functions continuous on $\mathcal{O}$.

Find sets $\{E_k\}$ for which is P injective, and if so find explicit formulas for its inverse.

The formulation of Problem (A) has appeared explicitly in a number of the papers cited above (see also [18] and [19]). Problem (B) as formulated here does not but is a generalization of the situation in [8], where the E_k are cubes in $\mathbf{R}^d$. The local Pompeiu problem when $m = 1$ (and all rotations are included) was first studied in [6] and its inversion, as well as the case when $m = 2$ and the E_k are balls in $\mathbf{R}^d$ was studied in [9]. Consideration of these problems relates to the more general problem of recovery of a function f on $\mathbf{R}^d$ from the convolution data $s_k = f * \mu_k$, where in the case of Problem (B), μ_k is the characteristic function of E_k. If (1) can be solved, then such an f can be recovered via

$$\sum_{k=1}^{m} s_k * v_k = \sum_{k=1}^{m}(f * \mu_k) * v_k = \sum_{k=1}^{m} f * (\mu_k * v_k)$$

$$= f * \sum_{k=1}^{m} \mu_k * v_k = f * \delta = f.$$

Clearly, inverting the operator P is sufficient to recover f from the data $\{s_k\}$ in the case of Problem (B). However, suppose that we wish to recover f in some compact set K containing U. If all we had at our disposal were compactly supported solutions to (1), then we would have to know the data s_k on the set $K + \operatorname{supp} v_k$. If the μ_k were characteristic functions of E_k, then this means that we could recover f on K from knowledge of its averages over shifts of the sets E_k that stay within the set $K + E_k + \operatorname{supp} v_k$. On the other hand, if we were in possession of a solution to Problem (B), then we could recover f on K from knowledge of its averages over shifts of the sets E_k that stay within only the set K. Therefore, both problems are a form of local deconvolution, with the second being in some sense more local than the first.

The goal of this paper is to show that solutions to Problem (A) when the μ_k are characteristic functions of compact sets E_k can lead naturally to solutions of Problem (B) by means of the theory of sampling and interpolation in Paley–Wiener spaces of entire functions.[2] It is shown, in particular, that the solution to Problem (A) permits deconvolution that is local in the strong sense of Problem (B). The connection is illustrated when the E_k are cubes in $\mathbf{R}^d$. This case is the only one in which the connection has been completely worked out in the literature (see [42, 22]). In that sense the results presented here are not new. However, intriguing possibilities exist for extensions to new situations.

---

[2] Given a compact set $E \subseteq \mathbf{R}^d$, the Paley–Wiener space $PW_E(\mathbf{R}^d)$ (sometimes simply PW_E) is defined as

$$PW_E(\mathbf{R}^d) = \{f \in L^2(\mathbf{R}^d) : \operatorname{supp} \widehat{f} \subseteq E\},$$

where $\widehat{f}$ is the Fourier transform of f. Given $b > 0$, $PW_{[-b,b]}(\mathbf{R})$ is denoted $PW_b(\mathbf{R})$. The Paley–Wiener theorem identifies PW_b with the space of functions entire in $\mathbf{C}$ of exponential type $2\pi b$ and square-integrable on $\mathbf{R}$. See [39, 28, 43, 44] for details.

2 Solutions of the multisensor deconvolution problem

We note first that the question of existence of solutions to (1) is classical and a characterization is given in the following theorem of Hörmander [24].

Theorem 1. *There exist compactly supported distributions* $\{v_k\}_{k=1}^m \subseteq \mathcal{E}'(\mathbf{R}^d)$ *solving* (1) *if and only if there exist constants* A, B, $N > 0$ *such that*

$$\sum_{k=1}^m |\widehat{\mu}_k(z)| \geq A(1+|z|)^{-N} e^{-B|\Im z|} \quad \forall z \in \mathbf{C}^d, \tag{3}$$

where $\widehat{\mu}_k$ *is the Fourier transform[3] of* μ_k.

The condition (3) is referred to as the *strongly coprime condition.*
    To handle the case of cubes in $\mathbf{R}^d$, it is sufficient to consider the case $d = 1$. The gist of the idea is there already in this case and the general case is obtained by simply taking cross products of one-dimensional solutions. See [42] for details. In this case we have that $E_k = [-r_k, r_k]$ and that $\mu_k = \chi_{[-r_k, r_k]}$. Note that for each k, the Fourier transform of μ_k is given by

$$\widehat{\mu}_k(z) = \frac{\sin(2\pi r_k z)}{\pi z},$$

that the zero set of $\widehat{\mu}_k(z)$ is precisely $\Lambda_k \equiv \{\frac{n}{2r_k} : n \in \mathbf{Z} \setminus \{0\}\}$, and that all of its zeros are simple. A result of Meisters and Petersen [37] shows that in this case the strongly coprime condition is equivalent to the ratios r_i/r_j being *poorly approximated by rationals*, that is, there exist numbers C, $N > 0$ such that for all integers $p, q, |r_i/r_j - (p/q)| \geq C|q|^{-N}$. We assume that this diophantine condition holds throughout this section.
    The first observation to make is that replacing δ in (1) by a function $\varphi \in C_c^\infty(\mathbf{R})$ with sufficiently small support allows both for the recovery of f by means of the computation of $f * \varphi$ for all such mollifiers φ, and also for the solution of (1) by letting $\varphi \to \delta$ in the sense of distributions. Specifically, we find solutions to the modified equation

$$\sum_{k=1}^m \mu_k * v_{k,\varphi} = \varphi \tag{4}$$

and then show that the solutions $v_{k,\varphi}$ so obtained each have a limit v_k in $\mathcal{E}'(\mathbf{R})$ as $\varphi \to \delta$ and that these limiting distributions satisfy (1). In this sense solving (4) is equivalent to solving (1). This question was considered when the E_k are intervals in [19] and was also studied in [36] when the E_k are cubes (as in [8]) or balls (as in [9]) and when the mollifiers φ are taken from an adapted polar wavelet basis.

---

[3] We define the Fourier transform of $f \in L^1(\mathbf{R}^d)$ by $\widehat{f}(z) = \int_{\mathbf{R}^d} f(t)e^{-2\pi i z \cdot t} dt$ for all $z \in \mathbf{C}^d$ for which the integral makes sense. For distributions f the integral is interpreted in the usual weak sense.

In the approach of Berenstein et al. to solving (4), the role of sampling theory is evident. Let $\varphi \in C_c^\infty(\mathbf{R})$ be supported in an interval of the form $[-A, A]$ with $A < R$. Consider first the generating function

$$G(z) = \prod_{k=1}^m \widehat{\mu}_k(z) = \prod_{k=1}^m \frac{\sin(2\pi r_k z)}{\pi z}.$$

Note first that $G(z)$ is an entire function of exponential type R, that is, there is a constant $C > 0$ such that for all $z \in \mathbf{C}$, $|G(z)| \leq Ce^{2\pi R|\Im(z)|}$. Next observe that since r_i/r_j is irrational for all $i \neq j$ (since each such ratio is poorly approximated by rationals), the sets Λ_k are mutually disjoint. Hence all the zeros of $G(z)$ are simple and lie at the points

$$\Lambda = \cup_{k=1}^m \Lambda_k = \bigcup_{k=1}^m \left\{ \frac{n}{2r_k} : n \in \mathbf{Z} \setminus \{0\} \right\}. \tag{5}$$

Next we construct a sequence of contours $\{\Gamma_n\}$ which we may take to be circles of radius ρ_n centered at the origin. The main observation is that we can choose the radii in such a way that $\inf_n \operatorname{dist}(\Gamma_n, \Lambda) > 0$, that is, the contours stay a fixed positive distance away from the zero set of $G(z)$. By basic properties of the sine function, this means that there is a $c > 0$ such that for all k,

$$|\sin(2\pi r_k z)| \geq ce^{2\pi r_k|\Im(z)|}. \tag{6}$$

Now we define for each n the function

$$I_n(z) = \frac{G(z)}{2\pi i} \int_{\Gamma_n} \frac{\varphi(\xi)}{(\xi - z)G(\xi)} d\xi.$$

By the residue theorem, for $|z| < \rho_n$,

$$I_n(z) = \frac{G(z)}{2\pi i} \int_{\Gamma_n} \frac{\widehat{\varphi}(\xi)}{(\xi - z)G(\xi)} d\xi$$

$$= G(z) \left(\frac{\widehat{\varphi}(z)}{G(z)} - \sum_{\zeta \in \Lambda, |\zeta| < \rho_n} \frac{\widehat{\varphi}(\zeta)}{(z - \zeta)G'(\zeta)} \right)$$

$$= \widehat{\varphi}(z) - \sum_{\zeta \in \Lambda, |\zeta| < \rho_n} \widehat{\varphi}(\zeta) \frac{G(z)}{G'(\zeta)(z - \zeta)}.$$

Now by standard arguments using the Paley–Wiener theorem and the estimate (6), it can be shown that the integral $I_n(z)$ goes to zero uniformly on compact subsets of $\mathbf{C}$ as n goes to infinity (for details, see, for example, [19, Lemma 9.1]). Hence we arrive at the formula

$$\widehat{\varphi}(z) = \lim_{n \to \infty} \sum_{\zeta \in \Lambda, |\zeta| < \rho_n} \widehat{\varphi}(\zeta) \frac{G(z)}{G'(\zeta)(z - \zeta)}$$

$$= \lim_{n \to \infty} \sum_{k=1}^{m} \sum_{\zeta \in \Lambda_k, |\zeta| < \rho_n} \widehat{\varphi}(\zeta) \frac{G(z)}{G'(\zeta)(z - \zeta)} \tag{7}$$

$$= \sum_{k=1}^{m} \prod_{j \neq k} \widehat{\mu}_j(z) \lim_{n \to \infty} \sum_{\zeta \in \Lambda_k, |\zeta| < \rho_n} \frac{\widehat{\varphi}(\zeta)}{G'(\zeta)} \frac{\widehat{\mu}_k(z)}{(z - \zeta)}.$$

Note that if $\zeta \in \Lambda_k$, then

$$G'(\zeta) = \widehat{\mu}'_k(\zeta) \prod_{j \neq k} \widehat{\mu}_j(\zeta) = \frac{2\pi r_k (-1)^k}{\pi \zeta} \prod_{j \neq k} \widehat{\mu}_j(\zeta).$$

The strongly coprime condition implies that

$$\prod_{j \neq k} \widehat{\mu}_j(\zeta) \geq C(1 + |\zeta|)^{-N(m-1)} \tag{8}$$

since if r_i / r_j is poorly approximated by rationals for each $i \neq j$, then the strongly coprime condition holds for each pair $\{\mu_i, \mu_j\}$. Hence for $\zeta \in \Lambda_i$, we can get directly from the definition of strongly coprime the estimate

$$\widehat{\mu_j}(\zeta) \geq C_{i,j} (1 + |\zeta|)^{-N}.$$

Combining these estimates gives (8) (see [42, Lemma 5]). Hence the sequence $\{G'(\zeta)^{-1}\}$ grows like a polynomial in ζ. Since $\varphi \in C_c^\infty(\mathbf{R})$, $\widehat{\varphi}(\zeta)$ decays faster than any polynomial in ζ. Hence the coefficients in the series

$$\sum_{\zeta \in \Lambda_k, |\zeta| < \rho_n} \frac{\widehat{\varphi}(\zeta)}{G'(\zeta)} \frac{\widehat{\mu}_k(z)}{(z - \zeta)}$$

decay rapidly and we conclude that the limit

$$\lim_{n \to \infty} \sum_{\zeta \in \Lambda_k, |\zeta| < \rho_n} \frac{\widehat{\varphi}(\zeta)}{G'(\zeta)} \frac{\widehat{\mu}_k(z)}{(z - \zeta)}$$

represents the Fourier transform of a distribution (in fact a continuous function) supported on the interval $[-r_k, r_k]$. Call this function $f_{k,\varphi}$.

Finally, note that since $m \geq 2$, the product $\prod_{j \neq k} \widehat{\mu}_j(z)$ is not empty for any k. If σ is a permutation of $\{1, 2, \ldots, m\}$ with the property that $\sigma(k) \neq k$ for all k, then in light of (7), we may write

$$\widehat{\varphi}(z) = \sum_{k=1}^{m} \prod_{j \neq k} \widehat{\mu}_j(z) \lim_{n \to \infty} \sum_{\zeta \in \Lambda_k, |\zeta| < \rho_n} \frac{\widehat{\varphi}(\zeta)}{G'(\zeta)} \frac{\widehat{\mu}_k(z)}{(z - \zeta)}$$

$$= \sum_{k=1}^{m} \prod_{j \neq \sigma(k)} \widehat{\mu}_j(z) \lim_{n \to \infty} \sum_{\zeta \in \Lambda_{\sigma(k)}, |\zeta| < \rho_n} \frac{\widehat{\varphi}(\zeta)}{G'(\zeta)} \frac{\widehat{\mu}_{\sigma(k)}(z)}{(z - \zeta)}$$

$$= \sum_{k=1}^{m} \widehat{\mu}_k(z) \prod_{j \neq k, \sigma(k)} \widehat{\mu}_j(z) \lim_{n \to \infty} \sum_{\zeta \in \Lambda_\sigma(k), |\zeta| < \rho_n} \frac{\widehat{\varphi}(\zeta)}{G'(\zeta)} \frac{\widehat{\mu}_{\sigma(k)}(z)}{(z - \zeta)}$$

$$= \sum_{k=1}^{m} \widehat{\mu}_k(z) \prod_{j \neq k, \sigma(k)} \widehat{\mu}_j(z) \widehat{f}_{\sigma(k), \varphi}(z)$$

$$= \sum_{k=1}^{m} \widehat{\mu}_k(z) \widehat{v}_{k, \varphi}(z),$$

and we have solved (4) with supp $v_{k,\varphi} \subseteq [-R + r_k, R - r_k]$ for each k.

The result of Berenstein and Yger that appears in [3] and [7] is considerably deeper and more general than the simplified version presented here. Also the following work on interpolating varieties should be noted in this context [13, 14, 15, 16, 30, 31, 32, 33].

3 A solution to the Local Pompeiu problem in the strongly coprime case

We have shown in the previous section that it is possible to represent any function $\varphi(x)$, C^∞ on $\mathbf{R}$ and supported in an interval of the form $[-A, A]$ with $A < R$ by means of an equation of the form

$$\varphi = \sum_{k=1}^{m} \mu_k * v_{k,\varphi}, \tag{9}$$

where the $v_{k,\varphi}$ are continuous functions supported in the interval $[-R + r_k, R - r_k]$. To see the connection to the solution of the Local Pompeiu Problem, we make the trivial observation that for any f on $\mathbf{R}$,

$$\int f(x)\varphi(-x)dx = f * \varphi(0)$$

$$= \sum_{k=1}^{m} f * \mu_k * v_{k,\varphi}(0)$$

$$= \sum_{k=1}^{m} \int f * \mu_k(x)v_{k,\varphi}(-x)dx$$

$$= \sum_{k=1}^{m} \int_{-R+r_k}^{R-r_k} f * \mu_k(x)v_{k,\varphi}(-x)dx.$$

Consequently, if $f * \mu_k$ vanishes on $[-R + r_k, R - r_k]$ for all k, then

$$\int f(x)\varphi(-x)dx = 0$$

for all C^∞ functions φ supported on closed subintervals of $(-R, R)$. Clearly, this is enough to guarantee that $f \equiv 0$ on $[-R, R]$. It also implies a generalization of Problem (B) to functions in $L^2([-R, R])$. In particular, we have proved the following theorem.

Theorem 2. *Let* $0 < r_1 < r_2 < \cdots < r_m$ *be such that the functions* $\{\chi_{[-r_k,r_k]}\}_{k=1}^m$ *form a strongly coprime system, and let* $R = \sum_k r_k$. *If for all* k, $f \in L^2([-R, R])$ *satisfies*

$$f * \mu_k(x) = 0 \quad \text{for all } x \in [-R + r_k, R - r_k],$$

then $f \equiv 0$.

Theorem 2 implies a solution to the special case of Problem (B) in which the convolvers satisfy the strongly coprime condition. Is it possible to remove this condition?

In order to answer this question, we note that the proof of Theorem 2 involved the following assertion: *there is a collection of functions* $\mathcal{F} \subseteq L^2([-R, R])$ *with the property that* (i) *each* $\varphi \in \mathcal{F}$ *can be represented in the form* (9) *and* (ii) *$\mathcal{F}$ is complete in* $L^2([-R, R])$. In our case, the collection $\mathcal{F}$ was the collection of all C^∞ functions supported on closed subintervals of $(-R, R)$.

The point in the proof where the strongly coprime condition was used was to control the decay to zero of the denominators $G'(\zeta)$, $\zeta \in \Lambda$, so that it could be matched by the decay of the numerators $\widehat{\varphi}(\zeta)$. Without the assumption of coprimality, the denominators in question could go to zero arbitrarily rapidly. The only way to counteract such rapid decay is to have numerators that actually vanish on all but finitely many of the points in Λ. This leads very naturally to the following questions: (i) *Do there exist interpolating functions* $\{\varphi_\lambda\}_{\lambda \in \Lambda}$ *in* $PW_R(\mathbf{R})$ *with the property that for* $\zeta \in \Lambda$, $\varphi_\lambda(\zeta) = 1$ *if* $\zeta = \lambda$ *and* 0 *otherwise?* (ii) *Is this collection complete in* $L^2([-R, R])$? This is precisely the problem of sampling and interpolation on the set Λ for functions in $PW_R(\mathbf{R})$.

4 Sampling and interpolation in Paley–Wiener spaces

What we are interested in is whether the set of interpolating functions corresponding to the discrete set Λ is complete in PW_R. It turns out that in our case, the answer is well known and classical. We begin with the notion of a basis and generalized basis in a Hilbert space.

Definition 1. *A sequence of vectors* $\{x_n\}_{n=1}^\infty$ *is* complete *in a Hilbert space* H *if* span($\{x_n\}$) *is dense in* H. *Equivalently,* $\{x_n\}_{n=1}^\infty$ *is* complete *in* H *if for any* $x \in H$, $\langle x, x_n \rangle = 0$ *for all* n *implies* $x = 0$.

A sequence $\{y_n\}_{n=1}^\infty$ *is* biorthogonal *to* $\{x_n\}_{n=1}^\infty$ *if* $\langle x_n, y_m \rangle = \delta_{n,m}$. *A sequence* $\{x_n\}_{n=1}^\infty$ *is* minimal *if it possesses a biorthogonal sequence and* exact *if it is both complete and minimal.*

A sequence of vectors $\{x_n\}_{n=1}^\infty$ *in a Hilbert space* H *is a* generalized basis *for* H *if there is a sequence* $\{n_j\}_{j=1}^\infty$ *of natural numbers such that* $n_j \to \infty$ *as* $j \to \infty$ *such that for each vector* $x \in H$ *there is a unique sequence of scalars* $\{a_n(x)\}_{n=1}^\infty$ *such that*

$$x = \lim_{j \to \infty} \sum_{n=1}^{n_j} a_n(x) x_n$$

in H. It is a basis *if it is a generalized basis for which* $n_j = j$.

The notion of a generalized basis (also referred to as a *basis with braces*) arose in the context of the theory of sampling and interpolation in Paley–Wiener spaces and appears in [26]. In that paper, B. Y. Levin introduces the notion of a sine-type function and shows that the zeros of such a function form a set of sampling and interpolation for Paley–Wiener spaces (Theorem 3 below) corresponding to a generalized basis for those spaces.

It is obvious that a generalized basis is complete, and it can be shown that a generalized basis is an exact sequence. Moreover, it can be shown that if $\{y_n\}_{n=1}^{\infty}$ is biorthogonal to a generalized basis $\{x_n\}$, then it is also a generalized basis (see [43, Theorem 5, p. 27] and [26]).

Since the Fourier transform is a unitary operator from $L^2([-R, R])$ onto the Paley–Wiener space PW_R, and in light of the fact that for any $\lambda \in \mathbf{C}$ and any $f \in L^2([-R, R])$,

$$\langle f, e^{2\pi i \lambda t} \rangle = \widehat{f}(\lambda),$$

the following dualities hold:

(a) A function $F \in PW_R$ is completely determined by its samples on a discrete set $\Lambda \subseteq \mathbf{C}$ if and only if the collection $\{e^{2\pi i \lambda t}\}_{\lambda \in \Lambda}$ is complete in $L^2([-R, R])$.

(b) There exists a collection of functions $\{\varphi_\lambda\}_{\lambda \in \Lambda} \subseteq PW_R$ interpolating $\Lambda \subseteq \mathbf{C}$ (that is, for $\lambda, \zeta \in \Lambda$, $\varphi_\lambda(\zeta) = 1$ if $\lambda = \zeta$ and 0 if $\lambda \neq \zeta$) if and only if the collection $\{e^{2\pi i \lambda t}\}_{\lambda \in \Lambda}$ is minimal in $L^2([-R, R])$.

(c) Let $\{\lambda_n\}_{n \geq 1}$ be a sequence of distinct numbers. There exists a sequence $\{\varphi_{\lambda_n}\}_{n \geq 1}$ interpolating $\{\lambda_n\}$ and a sequence $\{n_j\}_{j=1}^{\infty}$ going to infinity with j such that for all $F \in PW_R$, $F = \lim_{j \to \infty} \sum_{n=1}^{n_j} F(\lambda_n)\varphi_{\lambda_n}$ if and only if $\{e^{2\pi i \lambda_n t}\}_{n=1}^{\infty}$ is a generalized basis for $L^2([-R, R])$.

(d) We may take $n_j = j$ in (c) if and only if $\{e^{2\pi i \lambda_n t}\}_{n=1}^{\infty}$ is a basis for $L^2([-R, R])$.

In the case of (c), by defining $F_j = \{n_j, n_j + 1, \ldots, n_{j+1} - 1\}$, we can write

$$F = \sum_{j=1}^{\infty} \sum_{n \in F_j} F(\lambda_n)\varphi_{\lambda_n}.$$

We refer to this type of convergence as *block summation* which is weaker than conditional convergence.

The notion of a sine-type function was introduced by B. Y. Levin [26] advancing a method which appears already in [39] and [29] in which collections of complex exponentials $\{e^{2\pi i \lambda_n x}\}_{n \in \mathbf{Z}}$ are studied by examining functions of exponential type whose zero sets coincide exactly with the set $\{\lambda_n\}$. Such functions are called *generating functions* for the collection $\{e^{2\pi i \lambda_n x}\}$. Some fundamental resources for this approach are [23], [27], [28, Lecture 22-23], [40], and [25]. This point of view has been exploited to great effect in, for example, [20], [34], and [35].

Definition 2. *An entire function $F(z)$ is said to be of* sine type r *provided that*

(a) *for some $C > 0$, $|F(z)| \leq Ce^{2\pi r |\Im z|}$ for all $z \in \mathbf{C}$,*
(b) *if Γ is the zero set of $F(z)$, then $\sup_{\gamma \in \Gamma}\{|\Im \gamma|\} = H < \infty$, and*
(c) *for every $\epsilon > 0$, there is a positive constant c_ϵ such that whenever $\mathrm{dist}(z, \Gamma) > \epsilon$,*

$$c_\epsilon e^{2\pi r |\Im(z)|} \leq |F(z)|.$$

The main result of [26] is the following.

Theorem 3. *Let $F(z)$ be a function of sine type R with zero set $\{\gamma_n\}_{n=1}^\infty$ and suppose that each γ_n has multiplicity p_n. Then $\{t^j e^{2\pi i \gamma_n t}\}_{n=1, 0 \leq j < p_n}^\infty$ is a generalized basis for $L^2([-R, R])$.*

The generating function of interest to us is

$$S(z) = \prod_{k=1}^m \sin(2\pi r_k z),$$

which is a function of sine type R. $S(z)$ has a zero of order m at the origin and all of its other zeros are simple and occur on the set Λ. According to Theorem 3 and the remarks preceding it, we can write

$$F = \sum_{k=0}^{m-1} F^{(k)}(0)\varphi_{0,k} + \sum_{\lambda \in \Lambda} F(\lambda)\varphi_\lambda$$

for every $F \in PW_R$ where the last sum on the right-hand side is interpreted to converge by means of some block-summation criterion and where the functions $\varphi_{0,k}$ and φ_λ are interpolating functions. In this case, this means that for $0 \leq j, k < m$, $\varphi_{0,k}^{(j)}(0) = 1$ if $j = k$, 0 if $j \neq k$, and $\varphi_{0,k}(\lambda) = 0$ for all $\lambda \in \Lambda$; and that for $\lambda, \zeta \in \Lambda$, $\varphi_\lambda(\zeta) = 1$ if $\lambda = \zeta$, and 0 if $\lambda \neq \zeta$, and $\varphi_\lambda^{(j)}(0) = 0$ for $0 \leq j < m$.
    Levin gives explicit formulas for the interpolating functions which in this case are equivalent to the following. Define the meromorphic function $z \mapsto \Delta(z) = z^m/S(z)$ and for $0 \leq k < m$ define the entire functions

$$z \mapsto \varphi_{0,k}(z) = \sum_{j=k}^{m-1} \binom{j}{k} \Delta^{(j-k)}(0) \frac{S(z)}{z^{m-j}}$$

and for $\lambda \in \Lambda$,

$$z \mapsto \varphi_\lambda(z) = \frac{S(z)}{S'(\lambda)(z - \lambda)}.$$

Putting all of this together we have the following theorem.

Theorem 4. *Let $0 < r_1 < r_2 < \cdots < r_m$ be given such that r_i/r_j is irrational for $i \neq j$, let $R = \sum_{k=1}^m r_k$, and let Λ be given by (5). Then the set*

$$\{\varphi_{0,k}\}_{k=0}^{m-1} \cup \{\varphi_\lambda\}_{\lambda\in\Lambda}$$

defined above is a generalized basis for PW_R. This means that every $F \in PW_R$ can be written as

$$F = \sum_{k=0}^{m-1}\langle F, \varphi_{0,k}\rangle s_{0,k} + \sum_{\lambda\in\Lambda}\langle F, \varphi_\lambda\rangle s_\lambda,$$

where the sum over Λ is interpreted in the sense of block summation, and the functions $s_{0,k}$ and s_λ are the inverse Fourier transforms of the functions $(2\pi i t)^k \chi_{[-R,R]}$ and $e^{2\pi i \lambda t}\chi_{[-R,R]}$, respectively.

In order to connect Theorem 4 to the local Pompeiu problem, observe that each of the interpolating functions $\varphi_{0,k}$ and φ_λ contains a factor of the form $\sin(2\pi r_p z)/(\pi z)$ for some $1 \le p \le m$ possibly depending on λ. Specifically, note that

$$\varphi_{0,k}(z) = \frac{\sin(2\pi r_1 z)}{\pi z}\left(\pi\sum_{j=k}^{m-1}\binom{j}{k}\Delta^{(j-k)}(0)\frac{\prod_{\ell=2}^{m}\sin(2\pi r_\ell z)}{z^{m-j-1}}\right)$$

$$\equiv \frac{\sin(2\pi r_1 z)}{\pi z}H_{0,k}(z)$$

and that if $\lambda = n/2r_k$ for some $n \in \mathbf{Z}\setminus\{0\}$, then for any $p_\lambda \ne k$,

$$\varphi_\lambda(z) = \frac{\sin(2\pi r_{p_\lambda} z)}{\pi z}\left(\frac{\pi z\prod_{\ell\ne p_\lambda}\sin(2\pi r_\ell z)}{S'(\lambda)(z-\lambda)}\right)$$

$$\equiv \frac{\sin(2\pi r_{p_\lambda} z)}{\pi z}H_\lambda(z).$$

Here the functions $H_{0,k}$ and H_λ are entire functions representing the Fourier transforms of distributions supported in the intervals $[-R+r_1, R-r_1]$ and $[-R+r_{p_\lambda}, R-r_{p_\lambda}]$, respectively. Call these distributions $h_{0,k}$ and h_λ. Then given $F \in PW_R$ with inverse Fourier transform $f \in L^2([-R, R])$,

$$\langle F, \varphi_{0,k}\rangle = \left\langle F, \frac{\sin(2\pi r_1 x)}{\pi x}H_{0,k}\right\rangle = \left\langle F\frac{\sin(2\pi r_1 x)}{\pi x}, H_{0,k}\right\rangle = \langle f * \chi_{[-r_1,r_1]}, h_{0,k}\rangle$$

and

$$\langle F, \varphi_\lambda\rangle = \left\langle F, \frac{\sin(2\pi r_{p_\lambda} x)}{\pi x}H_\lambda\right\rangle = \left\langle F\frac{\sin(2\pi r_{p_\lambda} x)}{\pi x}, H_\lambda\right\rangle = \langle f * \chi_{[-r_{p_\lambda},r_{p_\lambda}]}, h_\lambda\rangle.$$

Finally, we can prove the following theorem.

Theorem 5. *Let $0 < r_1 < r_2 < \cdots < r_m$ be given such that r_i/r_j is irrational for $i \ne j$, let $R = \sum_{k=1}^{m} r_k$, and let Λ be given by (5). Then given $f \in L^2([-R, R])$, we can write*

$$f = \sum_{k=0}^{m-1} \langle f * \chi_{[-r_1,r_1]}, h_{0,k} \rangle (2\pi it)^k + \sum_{\lambda \in \Lambda} \langle f * \chi_{[-r_{p_\lambda}, r_{p_\lambda}]}, h_\lambda \rangle e^{2\pi i \lambda t}, \qquad (10)$$

where the $h_{0,k}$ and h_λ are distributions supported in $[-R + r_1, R - r_1]$ and $[-R + r_{p_\lambda}, R - r_{p_\lambda}]$, respectively, and where the sum over Λ converges in $L^2([-R, R])$ by block summation.

Theorem 5 is a generalization of Theorem 2 and gives not only uniqueness from local averages but also allows for the recovery of f from those local averages by means of an explicit nonharmonic Fourier series expansion.

5 Conclusions

The goal of this paper was to demonstrate how questions about sampling and interpolation of entire functions on $\mathbf{C}^d$ arise naturally from the work of Berenstein, Yger, et al. on local deconvolution in the specific case of local recovery of functions from their local averages.

(a) In the case of intervals in $\mathbf{R}$, we have shown that classical sampling theory for functions in the Paley–Wiener space PW_R provides an explicit inversion formula for the Pompeiu operator P, defined by (2). It is straightforward to generalize this result to cubes in $\mathbf{R}^d$ by taking tensor products of the interpolating functions $\varphi_{0,k}$ and φ_λ described in the previous section. Details may be found in [42] and [22].

(b) The inversion of the operator P given by (2) is in general unstable in the sense that the problem of finding f from the equation $Pf = g$ is an ill-posed inverse problem. In the case of intervals in $\mathbf{R}$ this instability is seen in the fact that the sum in (10) does not converge unconditionally (that is, regardless of the order in which the terms are summed) in $L^2([-R, R])$, nor does it converge conditionally, but by means of a block summation procedure arising from the result in Theorem 3. Unconditional convergence of the sum would imply the stable invertibility of P (in this case, the collection of interpolating functions would form a Riesz basis and not just a generalized basis). In any case, it can be shown that the sum over the *blocks* is unconditional. That is, for some partition of Λ into finite subsets $\{F_j\}$,

$$\sum_{j=1}^{\infty} \sum_{\lambda \in F_j} \langle f * \chi_{[-r_{p_\lambda}, r_{p_\lambda}]}, h_\lambda \rangle e^{2\pi i \lambda t}$$

converges independently of the order of the terms in j (see [41]). In this case, the interpolating functions form a *block Riesz basis* or a *Riesz basis from subspaces* [1, 21, 38]. This is stronger than a generalized basis.

(c) It is clear that the more general results of Berenstein, Yger, et al. do not lend themselves to such an easy interpretation in terms of sampling theory. However, it does seem that the techniques they employ can advance the theory of sampling and interpolation especially in higher dimensions. This may be a very fruitful and interesting line of research.

References

[1] S. Avdonin and S. Ivanov, *Families of Exponentials*, Cambridge University Press, Cambridge, UK, 1995.

[2] C. A. Berenstein and B. A. Taylor, The "three squares" theorem for continuous functions, *Arch. Rat. Mech. Anal.*, **63** (1977), 253–259.

[3] C. A. Berenstein and A. Yger, Le problème de la déconvolution, *J. Functional Anal.*, **54** (1983), 113–160.

[4] C. A. Berenstein, A. Yger, and B. A. Taylor, Sur quelques formules explicites de deconvolution, *J. Optics*, **14** (1983), 75–82.

[5] C. A. Berenstein and D. Struppa, Small degree solutions for the polynomial Bezout equation, *Linear Algebra Appl.*, **98** (1988), 41–55.

[6] C. A. Berenstein and R. Gay, Le Problème de Pompeiu local, *J. Anal. Math.*, **52** (1989), 133–166.

[7] C. A. Berenstein and A. Yger, Analytic Bezout identities, *Adv. Appl. Math.*, **10** (1989), 51–74.

[8] C. A. Berenstein, R. Gay, and A. Yger, The three squares theorem, a local version, in C. Sadosky, ed., *Analysis and Partial Differential Equations: A Collection of Papers Dedicated to Mischa Cotlar*, Lecture Notes in Pure and Applied Mathematics, Marcel Dekker, New York, 1990, 35–50.

[9] C. A. Berenstein and A. Yger, Inversion of the local Pompeiu transform, *J. Anal. Math.*, **54** (1990), 259–287.

[10] C. A. Berenstein and E. V. Patrick, Exact deconvolution for multiple convolution sensors: An overview plus performance characterizations for imaging sensors, *Proc. IEEE*, **78**-4 (1990) (special issue on multidimensional signal processing), 723–734.

[11] C. A. Berenstein, S. Casey, and E. V. Patrick, *Systems of Convolution Equations, Deconvolution, and Wavelet Analysis*, white paper, Systems Research Center, University of Maryland, College Park, MD, 1990.

[12] C. A. Berenstein and A. Yger, Effective Bezout identities in $Q[z_1, \ldots, z_n]$, *Acta Math.*, **166** (1991), 69–120.

[13] C. A. Berenstein and B. Q. Li, Interpolating varieties for weighted spaces of entire functions in $\mathbf{C}^n$, *Publ. Mat.*, **38**-1 (1994), 157–173.

[14] C. A. Berenstein and B. Q. Li, Interpolating varieties for spaces of meromorphic functions, *J. Geom. Anal.*, **5**-1 (1995), 1–48.

[15] C. A. Berenstein, B. Q. Li, and A. Vidras, Geometric characterization of interpolating varieties for the (FN)-space A_p^0 of entire functions, *Canad. J. Math.*, **47**-1 (1995), 28–43.

[16] C. A. Berenstein, D.-C. Chang, and B. Q. Li, Interpolating varieties and counting functions in $\mathbf{C}^n$, *Michigan Math. J.*, **42**-3 (1995), 419–434.

[17] C. A. Berenstein, J. Baras, and N. Sidiropolis, Two-dimensional signal deconvolution: Design issues related to a novel multisensor-based approach, in S.-S. Chen, *Stochastic and Neural Methods in Signal Processing, Image Processing, and Computer Vision*, Proceedings of SPIE 1569, SPIE, Bellingham, WA, 1991.

[18] S. Casey and D. Walnut, Systems of convolution equations, deconvolution, Shannon sampling and the Gabor and wavelet transform, *SIAM Rev.*, **36**-4 (1994), 537–577.

[19] S. Casey and D. Walnut, Residue and sampling techniques in deconvolution, in J. Benedetto and P. Ferreira, eds., *Modern Sampling Theory*, Birkhäuser, Boston, 2001, 193–218.

[20] K. M. Flornes, Y. Lyubarskii, and K. Seip, A direct interpolation method for irregular sampling, *Appl. Comput. Harmonic Anal.*, **7**-3 (1999), 305–314.

[21] I. Gohberg and M. Krein, *Introduction to the Theory of Linear Nonselfadjoint Operators*, Americal Mathematical Society, Providence, RI, 1969.

[22] K. Gröchenig, C. Heil, and D. Walnut, Nonperiodic sampling and the local three squares theorem, *Ark. Mat.*, **38** (2000), 77–92.

[23] J. R. Higgins, *Sampling Theory in Fourier and Signal Analysis*, Clarendon Press, Oxford, UK, 1996.

[24] L. Hörmander, Generators for some rings of analytic functions, *Bull. Amer. Math. Soc.*, **73** (1967), 943–949.

[25] S. V. Hruščev, N. K. Nikol'skiĭ, and B. S. Pavlov, Unconditional basees of exponentials and of reproducing kernels, in V. P. Havin and N. K. Nikol'skiĭ, eds., *Complex Analysis and Spectral Theory: Seminar, Leningrad*, 1979/80, Lecture Notes in Mathematics 864, Springer–Verlag, Berlin, New York, Heidelberg, 1981, 214–335.

[26] B. Y. Levin, On bases of exponential functions in L^2, *Zap. Mekh.-Mat. Fak. i Khar'kov. Mat. Obshch.*, **27** (1961), 39–48 (in Russian).

[27] B. Y. Levin, *Distribution of Zeros of Entire Functions*, American Mathematical Society, Providence, RI, 1980.

[28] B. Y. Levin, *Lectures on Entire Functions*, American Mathematical Society, Providence, RI, 1996.

[29] N. Levinson, *Gap and Density Theorems*, American Mathematical Society Colloquium Publications 26, American Mathematical Society, Providence, RI, 1940.

[30] B. Q. Li and B. A. Taylor, On the Bézout problem and area of interpolating varieties in $\mathbf{C}^n$, *Amer. J. Math.*, **118**-5 (1996), 989–1010.

[31] B. Q. Li, On the Bézout problem and area of interpolating varieties in $\mathbf{C}^n$ II, *Amer. J. Math.*, **120**-6 (1998), 1191–1198.

[32] B. Q. Li and E. Villamor, Interpolating multiplicity varieties in $\mathbf{C}^n$, *J. Geom. Anal.*, **11**-1 (2001), 91–101.

[33] B. Q. Li and E. Villamor, Interpolation in the unit ball of $\mathbf{C}^n$, *Israel J. Math.*, **123** (2001), 341–358.

[34] Y. Lyubarskii and K. Seip, Complete interpolating sequences for Paley–Wiener spaces and Muckenhoupt's (A_p) condition, *Rev. Mat. Iberoamericana*, **13**-2 (1997), 361–376.

[35] Y. Lyubarskii and K. Seip, Sampling and interpolating sequences for multiband-limited functions and exponential bases on disconnected sets, *J. Fourier Anal. Appl.*, **3**-5 (1997) (dedicated to the memory of Richard J. Duffin), 597–615.

[36] E. M. Maghras, Restitution des coefficients d'ondelette des signaux filtrés, *Colloq. Math.*, **68**-2 (1995), 265–283.

[37] G. Meisters and E. Petersen, Non-Liouville numbers and a theorem of Hörmander, *J. Functional Anal.*, **29** (1978), 142–150.

[38] N. K. Nikol'skiĭ, *Treatise on the Shift Operator*, Springer-Verlag, Berlin, New York, Heidelberg, 1985.

[39] R. E. A. C. Paley and N. Wiener, *Fourier Transforms in the Complex Domain*, American Mathematical Society, Providence, RI, 1934.

[40] B. S. Pavlov, Basicity of an exponential system and Muckenhoupt's condition, *Soviet Math. Dokl.*, **20**-4 (1979), 655–659.

[41] B. Rom and D. Walnut, Sampling and interpolation on unions of shifted lattices in one dimension, in C. Heil, ed., *Harmonic Analysis and Applications: A Volume in Honor of John Benedetto's 65th Birthday*, Birkhäuser, Boston, 2005.

[42] D. Walnut, Solutions to deconvolution equations using nonperiodic sampling, *J. Fourier Anal. Appl.*, **4**-6 (1998), 669–709.

[43] R. Young, *An Introduction to Nonharmonic Fourier Series*, 1st ed., Academic Press, New York, 1980.

[44] R. Young, *An Introduction to Nonharmonic Fourier Series*, revised 1st ed., Academic Press, New York, 2001.

[45] L. Zalcman, Analyticity and the Pompeiu problem, *Arch. Rational Mech. Anal.*, **47** (1972), 237–254.

[46] L. Zalcman, Offbeat integral geometry, *Amer. Math. Monthly*, **87** (1980, 161–175).

Orthogonal Projections on Hyperbolic Space

Yuying Qiao[1] and John Ryan[2]

[1] Department of Mathematics
    Hebei Normal University
    Shijiazhuang
    People's Republic of China
    yuyingqiao@163.com
[2] Department of Mathematics
    University of Arkansas
    Fayetteville, AR 72701
    USA
    jryan@uark.edu

Dedicated to Carlos Berenstein on the occasion of his 60*th birthday.*

Summary. A well-known decomposition of the L^2 space of a bounded domain in the complex plane is extended here to the context of hyperbolic n-space. We will use the model of upper half-space with the hyperbolic metric. Applications to boundary value problems for the hyperbolic Laplacian and another Laplace operator are indicated.

1 Introduction

Function theory for Dirac operators on manifolds have been developed in [C, CC, Cn, M]. In the meantime, in a number of papers including [E, EL, EL2, EL1, L] function theory for a hyperbolic Dirac–Hodge operator on upper half-n-space endowed with the hyperbolic metric has been developed. In [Q], the authors, following [E], develop much of the basic function theory for this operator and the hyperbolic Laplace operator.

In this paper, we continue this analysis by showing that a well-known decomposition of L^2 integrable functions on a bounded domain in the complex plane in terms of Bergman spaces and Sobolev spaces carry through to the context described here. This may be applied to solving boundary value problems described in [GS] and elsewhere.

2 Preliminaries

Here we will consider upper half-space $R^{n,+}$ endowed with the hyperbolic metric $ds^2 = \frac{dx_1^2 + \cdots + dx_n^2}{x_n^2}$. With respect to this metric, one may consider the adjoint δ to the de Rham exterior derivative d. Namely, $\delta = \star d \star$, where $\star$ is the Hodge star map acting on sections in the alternating bundle over R^{n+}. The Dirac–Hodge operator is the differential operator $d + \delta$ acting on differentiable sections on the alternating algebra $\Lambda(R^{n,+})$. The square of $d + \delta$ is the Laplacian $d\delta + \delta d$ with respect to the hyperbolic or Poincaré metric. To better understand the Dirac–Hodge operator, let us first follow [L] and note that as a vector space the alternating or exterior algebra $\Lambda(R^n)$ is isomorphic to the Clifford algebra Cl_n generated from R^n with negative definite inner product. Namely, let us consider $\mathbf{R}^n$ with orthogonal basis $e := \{e_1, \ldots, e_n\}$. Then Cl_n has as its basis

$$1, e_1, \ldots, e_n; e_1 e_2, \ldots, e_{n-1} e_n; \ldots; e_1 \ldots, e_n$$

and $e_i e_j + e_j e_i = -2\delta_{ij}$.

We may express the Clifford algebra as $Cl_n = Cl_{n-1} + Cl_{n-1} e_n$, where Cl_{n-1} is the Clifford algebra generated from R^{n-1} with orthonormal basis $e_1, \ldots, e_{n-1}$. Thus if $A \in Cl_n$, there are unique elements B and $C \in Cl_{n-1}$ such that $A = B + C e_n$. This gives rise to a pair of maps

$$P : Cl_n \to Cl_{n-1} : P(A) = B$$

and

$$Q : Cl_n \to Cl_{n-1} : Q(A) = C.$$

We will denote $-e_n Q(A) e_n \in Cl_{n-1}$ by $Q'(A)$.

The Dirac–Hodge operator, $d + \delta$, acting on vector-valued sections, now re-translates in Clifford algebra notation as $D + \frac{n-2}{x_n} Q'$, where $D = \Sigma_{j=1}^{n} e_j \frac{\partial}{\partial x_j}$ is the Euclidean Dirac operator. Thus the hyperbolic Dirac equation is defined to be $Df + \frac{n-2}{x_n} Q'(f) = 0$, where $f : U \to Cl_n$ is a differentiable function and U is a domain in $R^{n+} = \{x = x_1 e_1 + \cdots = x_n e_n : x_n > 0\}$. See [EL2] for more details. We shall abbreviate this Dirac equation to $Mf = 0$.

Note [EL] that if U is a domain in upper half-space and $h : U \to Cl_n$ is a C^2 function, then

$$-M^2 h = \triangle P(h) - \frac{n-2}{x_n} \frac{\partial P(h)}{\partial x_n} + \left(\triangle Q(h) - \frac{n-2}{x_n} \frac{\partial Q(h)}{\partial x_n} + \frac{n-2}{x_n^2} Q(h) \right) e_n,$$

where $\triangle$ is the Euclidean Laplacian.

In [A], it is noted for any real-valued function $u(x)$ defined on the domain U that $\triangle u - \frac{n-2}{x_n} \frac{\partial u}{\partial x_n}$ is the Laplace, or Laplace–Beltrami, formula for upper half-space with respect to the hyperbolic metric. We will denote this Laplacian by $\triangle_{R^{n,+}}$. We will call a Cl_{n-1}-valued solution to the equation $\triangle h - \frac{n-2}{x_n} \frac{\partial u}{\partial x_n} = 0$ a hyperbolic harmonic function. It follows that if f is hypermonogenic and C^2, then $P(f)$ is hyperbolic harmonic. Furthermore, we shall denote the operator $\triangle - \frac{n-2}{x_n} \frac{\partial}{\partial x_n} + \frac{n-2}{x_n^2}$ by $\triangle_{R^{n,+}}^\star$. The equations $\triangle_{R^{n,+}} u = 0$ and $\triangle_{R^{n,+}}^\star u = 0$ are both examples of the Weinstein equation. See, for instance, [Al, GSi, H, L1, W] for details.

Taking $A = a_0 + \cdots + a_{1\cdots n}e_1 \cdots e_n \in Cl_n$, we define the norm of A to be, as usual, $\|A\| = (a_0^2 + \cdots + a_{1\ldots n}^2)^{\frac{1}{2}}$. Using the antiautomorphism $- : Cl_n \to Cl_n :$ $-(e_{j_1} \cdots e_{j_r}) = -1^r e_{j_r} \cdots e_{j_1}$ it may be seen that $\|A\|^2$ is the real part of $A\overline{A}$, where $\overline{A}$ is defined to be $-(A)$.

3 Some integral formulas

Following [A], it may be seen that $G(x, y) = \int_{\frac{\|x-y\|}{\|x-\hat{y}\|}}^{1} \frac{(1-t^2)^{n-2}}{t^{n-1}} dt$ is a hyperbolic harmonic function. As $G(x, y)$ is real valued, then trivially $Q(G(x, y)) = 0$. Consequently, $MG(x, y) = DG(x, y)$. Therefore [L], the function $p(x, y) = DG(x, y)$ is a vector-valued hypermonogenic function. Here M and D are acting with respect to the x variable. Following [E, EL2], it may be noted that

$$DG(x, y) = \frac{(1 - s^2)^{n-2}}{s^{n-1}} \Big|_{\frac{\|x-y\|}{\|x-\hat{y}\|}}^{1} D\frac{\|x - y\|}{\|x - \hat{y}\|}$$
$$= x_n^{n-2} y_n^{n-1} \left(\frac{(x - y)}{\|x - y\|^n} e_n \frac{(x - \hat{y})}{\|x - \hat{y}\|^n} \right),$$

where $\hat{y} = y_1 e_1 + \cdots + y_{n-1}e_{n-1} - y_n e_n$.

Now suppose that U is a domain in upper half-space, and for two C^1 functions f and g defined on U and taking values in Cl_n, we consider the integral $\int_S g(x) \frac{n(x)}{x_n^{n-2}} f(x) d\sigma(x)$, where S is a compact smooth hypersurface lying in U, $n(x)$ is the outer unit normal vector to S at x and σ is the Lebesgue surface measure of S. On assuming that S is the boundary of a bounded subdomain V of U, then on applying Stokes' theorem we obtain

$$\int_S g(x) \frac{n(x)}{x_n^{n-2}} f(x) d\sigma(x) = \int_V \Big((g(x)D) \frac{1}{x_n^{n-2}} f(x) + g(x) \frac{1}{x_n^{n-2}} Df(x) $$
$$- g(x) \frac{(n - 2)}{x_n^{n-1}} e_n f(x) \Big) dx^n.$$

The next lemma then follows.

Lemma 1 ([EL2]). *Suppose f, g, U, S, and V are as in the previous paragraph. Then*

$$P \int_S \left(g(x) \frac{n(x)}{x_n^{n-2}} f(x) d\sigma(x) \right) = \int_V P[(g(x)M)) f(x) + g(x)(Mf(x))] \frac{dx^n}{x_n^{n-2}}.$$

Consequently, if $Mf = 0$ and $gM = 0$, where $gM = \Sigma_{j=1}^{n} \frac{\partial g(x)}{\partial x_j} e_j + \frac{n-2}{x_n} Q(g)$, we have the version of Cauchy's theorem established in [EL2]. Namely,

$$P \left(\int_S g(x) \frac{n(x)}{x_n^{n-2}} f(x) d\sigma(x) \right) = 0.$$

It may now be determined that for each $y \in V$,

$$P(f(y)) = \frac{1}{\omega_n} P\left(\int_S p(x, y) \frac{n(x)}{x_n^{n-2}} f(x) d\sigma(x)\right).$$

This is the Cauchy integral formula arising in [EL2]. Now let us consider $D_y G(x, y)$, where $D_y = \sum_{j=1}^n e_j \frac{\partial}{\partial y_j}$. As $\|x - \hat{y}\| = \|y - \hat{x}\|$, then $G(x, y)$ is hyperbolic harmonic in both the variables x and y, and

$$D_y G(x, y) = D_y \int_{\frac{\|y-x\|}{\|y-\hat{x}\|}} \frac{(1-s^2)^{n-2}}{s^{n-1}} ds = h(x, y) = p(y, x)$$

is hypermonogenic in the variable y.

In [Q] and elsewhere, it is noted that the space of solutions to the equation $Mf = 0$ on a domain U is a right module with respect to the algebra Cl_{n-1}. Thus if f is such that $Mf(x) = 0$ on U and $A \in Cl_{n-1}$, then $Mf(x)A = 0$ also on U. Now suppose that $r(x, y)$ is a Cl_n-valued function defined on a $U \times V$, where U and V are domains in $R^{n,+}$ and $r(x, y)$ is hypermonogenic in the variable y. Let M_y denote the hyperbolic Dirac operator with respect to the variable y. Then for any integrable function $g : U \to Cl_{n-1}$, we have $M_y \int_U r(x, y) g(x) dx^n = 0$. It now follows from simple adaptations of classical arguments from one-variable complex analysis that we have the following theorem.

Theorem 1 ([Q]). *Suppose that ψ is a Cl_{n-1}-valued C^∞ function with compact support on upper half-space. Then*

$$M_y \frac{1}{\omega_n} \int_{R^{n,+}} h(x, y) \psi(x) \frac{dx^n}{x_n^{n-2}} = \psi(y).$$

In [E], the kernel $q(x, y) = DH(x, y)$ is introduced, where

$$H(x, y) = \frac{1}{(n-2)} \frac{1}{\|x - y\|^{n-2}\|x - \hat{y}\|^{n-2}}$$

$$= \frac{1}{(n-2)} \frac{1}{\|x - y\|^{n-2}\|y - \hat{x}\|^{n-2}}.$$

In [E], it is shown that the kernel $q(x, y)$ is the Cauchy kernel for the Q part of a Cauchy integral formula for hypermonogenic functions.

From [E], we have

$$f(y) = P(f(y)) + Q(f(y))e_n$$

$$= \frac{2^{n-1} y_n^{n-2}}{\omega_n} \left(P\left(\int_{\partial U} r(x, y) \frac{n(x)}{x_n^{n-2}} f(x)\right) d\sigma(x)\right)$$

$$- Q\left(\int_{\partial U} q(x, y) n(x) f(x)) d\sigma(x)\right) e_n,$$

where $r(x, y) = y_n^{-n+2} p(x, y)$.

Let us denote the kernel $D_y H(x, y)$ as $s(x, y)$. Following [E], we have that $D_y s(x, y) - \frac{n-2}{y_n} Q'(s(x, y)) = 0$ and $y_n^{n-2} s(x, y) e_n$ is hypermonogenic in the variable y. One may determine the following.

Proposition 1. *Suppose* $\psi : U \to Cl_{n-1}$ *is a* C^1 *function. Then for each* $y \in U$,

$$\psi(y) = -\left(M_y \frac{2^{n-2}}{\omega_n} y_n^{n-2} \int_U s(x, y) e_n \psi(x) dx^n \right) e_n.$$

Now for any $A \in Cl_n$, $P(A) = \frac{1}{2}(A + \hat{A})$, where $\hat{A} = B - Ce_n$ with B and $C \in Cl_{n-1}$. Moreover, $Q(A) = \frac{-e_n}{2}(A - \hat{A})$ and for any elements X and $Y \in Cl_n$ it is straightforward to determine that $\hat{XY} = \hat{X}\hat{Y}$. Using these observations, it is noted in [E] that the Cauchy integral formula becomes

$$f(y) = \frac{1}{\omega_n} 2^{n-1} y_n^{n-2} \left(\int_{\partial U} \frac{1}{2}(r(x, y) n(x) \frac{n(x)}{x_n^{n-2}} f(x) + \hat{r}(x, y) \frac{\hat{n}(x)}{x_n^{n-2}} \hat{f}(x) \right) d\sigma(x)$$
$$- \frac{e_n}{2}(q(x, y) n(x) f(x) - \hat{q}(x, y) \hat{n}(x) \hat{f}(x)) d\sigma(x)).$$

In [E], it is shown that this expression simplifies to

$$f(y) = \frac{2^{n-1} y_n^{n-2}}{\omega_n} \left(\int_{\partial U} \frac{(x - y)^{-1} n(x) f(x)}{\|x - y\|^{n-2} \|x - \hat{y}\|^{n-2}} d\sigma(x) \right.$$
$$\left. - \int_{\partial U} \frac{(\hat{x} - y)^{-1} n(\hat{x}) \hat{f}(x)}{\|x - y\|^{n-2} \|\hat{x} - y\|^{n-2}} d\sigma(x) \right).$$

If we write $E(x, y)$ for $\frac{(x-y)^{-1}}{\|x-y\|^{n-2} \|x-\hat{y}\|^{n-2}}$ and $F(x, y)$ for $\frac{(\hat{x}-y)^{-1}}{\|x-y\|^{n-2} \|\hat{x}-y\|^{n-2}}$, then this integral formula simplifies to

$$f(y) = \frac{2^{n-1} y_n^{n-2}}{\omega_n} \int_{\partial U} (E(x, y) n(x) f(x) - F(x, y) \hat{n}(x) \hat{f}(x)) d\sigma(x).$$

In [Q], the following is shown.

Theorem 2. *Suppose that* $\psi \in C^1(\overline{U})$ *and* $y \in U$. *Then*

$$\psi(y) = M_y y_n^{n-2} \left(\int_U E(x, y) \psi(x) dx^n - \int_U F(x, y) \hat{\psi}(x) dx^n \right).$$

The following is also shown in [Q].

Theorem 3. *Suppose that* S *is a compact, strongly Lipschitz surface lying in upper half-space. Suppose also that* S *is the boundary of a bounded domain* U^+ *and an exterior domain* $U^- \subset R^{N,+}$. *Then for each function* $\phi \in L^p(S)$ *for* $1 < p < \infty$ *and for a path* $y_\pm(t) \in U^\pm$ *with nontangential limit* $y(1) = y \in S$, *we have*

$$\lim_{t \to 1} \frac{2^{n-2} y_\pm(t)^{n-2}}{\omega_n} \int_S (E(x, y(t))n(x)\phi(x) - F(x, y(t))\hat{n}(x)\hat{\phi}(x))d\sigma(x)$$

$$= \pm \frac{1}{2}\phi(y) + \frac{2^{n-2}}{\omega_n} PV \int_S y_n^{n-2}(E(x, y)n(x)\phi(x) - F(x, y)\hat{n}(x)\hat{\phi}(x))d\sigma(x)$$

for almost all $y \in S$.

We shall denote the singular integral

$$\frac{2^{n-2} y_n^{n-2}}{\omega_n} PV \int_S (E(x, y)n(x)\psi(x) - F(x, y)\hat{n}(x)\hat{\psi}(x))d\sigma(x)$$

by $T_S(\psi)$.

A minor adaptation of the proof of [E, Theorem 17] tells us the following.

Theorem 4 ([Q]). *Suppose S is a Lipschitz surface lying in the closure of upper half-space and $\phi \in L^p(S)$ for some $p \in (1, \infty)$. Then the integral*

$$\frac{2^{n-2} y_n^{n-2}}{\omega_n} \int_S (E(x, y)n(x)\phi(x) - F(x, y)\hat{n}(x)\hat{\phi}(x))d\sigma(x)$$

defines a left-hypermonogenic function $f(y)$ on $R^{n,+} \backslash S$.

As $\lim_{y_n \to \infty} y_n^{n-2} E(x, y) = 0$ and $\lim_{y_n \to \infty} y_n^{n-2} F(x, y) = 0$ for each $x \in S$ and $\lim_{y_n \to 0} y_n^{n-2} E(x, y) = \lim_{y_n \to 0} y_n^{n-2} F(x, y) = 0$ for each $x \in S$, then $\lim_{y_n \to \infty} f(y) = \lim_{y_n \to 0} f(y) = 0$. It now follows that the operators $\frac{1}{2}I \pm T_S :$ $L^p(S) \to L^p(S)$ are projection operators with images the Hardy spaces $H^p(U^\pm) = \{f : U^\pm \to Cl_n : f$ is left hypermonogenic and nontangentially approaches some element in $L^p(S)\}$.

Consequently,

$$L^p(S) = H^p(U^+) \oplus H^p(U^-).$$

The operators $\frac{1}{2} \pm T_S$ are generalizations of the Plemelj projection operators to the context of hypermonogenic functions.

4 Orthogonal projections

Let us consider a bounded domain U with strongly Lipschitz boundary ∂U. Further, U and its boundary lie in upper half-space. Let $L^2(U, Cl_{n-1})$ denote the space of L^2 integrable, Cl_{n-1}-valued functions defined on U. We will first work with the inner product $P(\int_U f(x)\overline{g}(x)\frac{dx^n}{x_n^{n-2}})$. We want a description of the subspace of $L^2(U, Cl_{n-1})$ consisting of all those functions that are orthogonal to square integrable, right-hypermonogenic functions with respect to this inner product. Clearly, for each y belonging to the complement of $U \cup \partial U$, the function $p(x, y)$ is square integrable over U and is right hypermonogenic in x. Thus if ψ is orthogonal with respect to $\langle , \rangle$ to all square integrable hypermonogenic functions defined on U, then $\frac{1}{\omega_n} \int_U p(x, y)\overline{\psi}(x)\frac{dx^n}{x_n^{n-2}} = 0$.

However, from Theorem 1, we have

$$\overline{\psi}(w) = M_w \frac{1}{\omega_n} \int_U p(w, x)\overline{\psi}(x)\frac{dx^n}{x_n^{n-2}}.$$

Thus

$$0 = \left(\frac{1}{\omega_n}\right)^2 \int_U p(w, y) M_w \int_U p(w, x)\overline{\psi}(x)\frac{dx^n}{x_n^{n-2}}\frac{dw^n}{w_n^{n-2}}.$$

By Lemma 1, this gives us

$$\left(\frac{1}{\omega_n}\right)^2 P\left(\int_{\partial U} p(w, y)\frac{n(w)}{w_n^{n-2}} \int_U p(w, x)\overline{\psi}(x)\frac{dx^n}{x_n^{n-2}}d\sigma(w)\right) = 0.$$

We will denote the function $\frac{1}{\omega_n} \int_U p(w, x)\overline{\psi}(x)\frac{dx^n}{x_n^{n-2}}$ by $g(w)$. Thus we have

$$\frac{1}{\omega_n} P\left(\int_{\partial U} p(w, y)\frac{n(w)}{w_n^{n-2}}g(w)d\sigma(w)\right) = 0$$

for each y in the complement of $U \cup \partial U$. Consequently, if we allow y to move along a path which approaches ∂U nontangentially, we have that $P(-\frac{1}{2}g(y) + \frac{1}{\omega_n}P.V.\int_{\partial U} p(w, y)\frac{n(w)}{w_n^{n-2}}g(w)d\sigma(w)) = 0$ for almost all $y \in \partial U$.

Now let us introduce the function

$$f_g(y) = P\left(\frac{1}{\omega_n} \int_{\partial U} p(w, y)\frac{n(w)}{w_n^{n-2}}g(w)d\sigma(w)\right)$$

defined on U. Clearly, as y moves along a path that approaches the boundary of U nontangentially, we get as boundary value of $f_g(y)$ the expression

$$\frac{1}{2}f_g(y) + P\left(\frac{1}{\omega_n}P.V.\int_{\partial U} p(w, y)\frac{n(w)}{w_n^{n-2}}f_g(w)d\sigma(w)\right)$$

almost everywhere. Consequently, the function $g(x) - f_g(x)$ has the property that $P(g(x) - f_g(x)) = 0$ on ∂U.

Following Proposition 1, we can now use similar arguments to those we have just used to obtain the following.

Lemma 2. *Suppose that* $\theta : U \to e_n Cl_{n-1}$ *is a* C^1 *function, then for each* $y \in U$

$$\overline{\theta}(y) = -Q\left(M_y \frac{y_n^{n-2}}{\omega_n} \int_U s(x, y)\overline{\theta}(x)dx^n\right)e_n.$$

Thus

$$0 = Q\left(\left(\frac{1}{\omega_n}\right)^2 \int_U s(x, y) M_y y_n^{n-2} \int_U s(w, y)\overline{\theta}(w)dw^n dy^n\right).$$

Consequently,

$$0 = Q\left(\frac{1}{\omega_n} \int_{\partial U} s(x, y)n(y)h(y)d\sigma(y)\right),$$

where

$$h(y) = \frac{y_n^{n-2}}{\omega_n} \int_U s(w, y)\overline{\theta}(w)dw^n.$$

Let $F_h(y) = Q(\frac{y_n^{n-2}}{\omega_n} \int_{\partial U} s(w, y)n(w)h(w)dw^n)$.

Consequently, the function $h(x) - F_h(x)$ has the property that $Q(h - F_h) = 0$ almost everywhere on ∂U. Thus $h(x) - F_h$ has the property that it is orthogonal to any L^2 hypermonogenic function defined on U with respect to the inner product $Q(\int_U f(x)\overline{g}(x)dx^n)$.

The previous calculations leads us to define an inner product $\langle f, g \rangle$ for square integrable functions f and g taking values in Cl_n to be

$$\int_U P(f(x))P(\overline{g}(x))\frac{dx^n}{x_n^{n-2}} + \int_U Q(f(x))Q(\overline{g}(x))dx^n.$$

We shall denote this inner product space by $L^2_{\langle,\rangle}(U, Cl_n)$. Further, the Bergman space $\{f : U \to Cl_n : fM = 0 \text{ and } f \in L^2_{\langle,\rangle}(U, Cl_n)\}$ is denoted by $B_{\langle,\rangle}(U)$.

From Theorem 4, we know that $P(f_g)(x) + Q(F_h)(x)e_n$ is a hypermonogenic function on U. We denote this function by $\Theta(\mu)$, where $\mu(x) = \psi(x) + \theta(x)$.

Therefore, if $f(x) \in L^2_{\langle,\rangle}(U, Cl_n)$ is such that $\langle f, g \rangle = 0$ for each $g \in B_{\langle,\rangle}(U)$, then for each y in the complement of $U \cup \partial U$, we have

$$\frac{2^{n-1}y_n^{2-n}}{\omega_n}\left(P\left(\int_{\partial U} r(x, y)\frac{n(x)}{x_n^{n-2}}\psi(x)d\sigma(x)\right)\right.$$
$$\left. - Q\left(\int_{\partial U} q(x, y)n(x)\psi(x)\right)d\sigma(x)\right) = 0, \tag{1}$$

where $r(x, y) = y_n^{2-n}p(x, y)$ and

$$\psi(y) = y_n^{n-2}\left(\int_U E(x, y)f(x)dx^{n-2} - \int_U F(x, y)\hat{f}(x)dx^n\right).$$

Following arguments given in [E], it may be seen that equation (1) simplifies to

$$\frac{2^{n-1}y_n^{n-2}}{\omega_n}\int_{\partial U}(E(x, y)n(x)\psi(x) - F(x, y)\hat{n}(x)\hat{\psi}(x))d\sigma(x) = 0.$$

Now if we let y approach the boundary of U nontangentially, we get from the Plemelj formulas that

$$\frac{1}{2}\psi(y) = \frac{2^{n-1}y_n^{n-2}}{\omega_n}P.V.\int_{\partial U}(E(x,y)n(x)\psi(x) - F(x,y)\hat{n}(x)\hat{\psi}(x))d\sigma(x)$$

almost everywhere on ∂U. Consequently,

$$\psi(w) - \frac{2^{n-1}w_n^{n-2}}{\omega_n}\int_{\partial U}E(x,w)n(x)\psi(x) - F(x,w)\hat{n}(x)\hat{\psi}(x))d\sigma(x)$$

defines a function Ψ on U satisfying

 (i) $\Psi(w) = 0$ on ∂U;
(ii) $M\Psi(w) = f(w)$.

So we have established the following.

Theorem 5.
$$L^2(U, Cl_n) = B_{(,)}(U) \oplus MH(U),$$

where $H(U) = \{r : U \to Cl_n : r|_{\partial U} = 0 \text{ and } Mr \in L^2(U, Cl_n)\}$.

It follows that the methods described in [GS] for tackling boundary value problems can now be readily adapted to the context described here. In particular, one can now handle them in a similar way to those methods used in [GS] boundary problems for the operators $\triangle_{R^n,+}$ and $\triangle_{R^n,+}^\star$. First, in order to solve the equation $\triangle_{R^n,+}u = f$, where f is a bounded, Cl_{n-1}-valued C^1 function defined on a bounded domain U, one solution is given by $u = P(I^2 f)$, where

$$I(f)(y) = y_n^{n-2}\int_U(E(x,y)f(x) - F(x,y)\hat{f}(x))dx^n.$$

See [Q] for details. If we impose the further condition that $u|_{\partial U} = 0$, then we simply apply the constructions obtained in this section to produce the solution $P(I^2(f)) - \Theta(I(f))$. This is in complete analogy to arguments developed in [GS] for the Euclidean Laplacian.

References

[A]   L. V. Ahlfors, *Möbius Transformations in Several Dimensions*, Ordway Lecture Notes, University of Minnesota, Minneapolis, 1981.

[Al]   Ö. Akin and H. Leutwiler, On the invariance of the solutions of the Weinstein equation under Möbius transformations, in K. Gowrisankran, J. Bliedtner, D. Feyel, M. Goldstein, W. K. Hayman, and I. Netuka, eds., *Classical and Modern Potential Theory and Applications: Proceedings of the NATO Advanced Research Workshop, Chateau De Bonas, France, July 25–31, 1993*, Kluwer, Dordrecht, the Netherlands, 1994, 19–29.

[C]   D. Calderbank, *Dirac Operators and Clifford Analysis on Manifolds*, Preprint 96-131, Max Plank Institute for Mathematics, Bonn, 1996.

[CC]   P. Cerejeiras and J. Cnops, Hodge–Dirac operators for hyperbolic space, *Complex Variables*, **41** (2000), 267–278.

[Cn]   J. Cnops, *An Introduction to Dirac Operators on Manifolds*, Progress in Mathematical Physics, Birkhäuser, Boston, 2002.

[E]   S.-L. Eriksson, Integral formulas for hypermonogenic functions, to appear.

[EL]   S.-L. Eriksson-Bique and H. Leutwiler, Hypermonogenic functions, in J. Ryan and W. Spröβig, eds., *Clifford Algebras and Their Applications in Mathematical Physics*, Vol. 2, Birkhäuser, Boston, 2000, 287–302.

[EL1]   S.-L. Eriksson and H. Leutwiler, *k*-hypermonogenic functions, to appear.

[EL2]   S.-L. Eriksson and H. Leutwiler, Some integral formulas for hypermonogenic functions, to appear.

[EL3]   S.-L. Eriksson and H. Leutwiler, Hypermonogenic functions and their Cauchy-type theorems, in *Trends in Mathematics: Advances in Analysis and Geometry*, Birkhäuser, Basel, 2003, 97–112.

[GSi]   K. Gowrisankran and D. Singman, Minimal fine limits for a class of potentials, *Potential Anal.*, **13** (2000), 103–114.

[GS]   K. Gürlebeck and W. Sprössig, *Quaternionic Analysis and Elliptic Boundary Value Problems*, Birkhäuser, Basel, 1990.

[H]   A. Huber, On the uniqueness of generalized axially symmetric potentials, *Ann. Math.*, **60** (1954), 351–358.

[L]   H. Leutwiler, Modified Clifford analysis, *Complex Variables*, **17** (1992), 153–171.

[L1]   H. Leutwiler, Best constants in the Harnak inequality for the Weinstein equation, *Aequationes Math.*, **34** (1987), 304–315.

[M]   M. Mitrea, Generalized Dirac operators on non-smooth manifolds and Maxwell's equations, *J. Fourier Anal. Appl.*, **7** (2001), 207–256.

[Q]   Y. Qiao, S. Bernstein, S.-L. Eriksson, and J. Ryan, Function theory for Laplace and Dirac–Hodge operators in hyperbolic space, to appear.

[W]   A. Weinstein, Generalized axially symmetric potential theory, *Bull. Amer. Math. Soc.*, **59** (1953), 20–38.

Eigenwavelets of the Wave Equation

Gerald Kaiser

Signals and Waves
3803 Tonkawa Trail #2
Austin, TX 78756
USA
www.wavelets.com
kaiser@wavelets.com

To Carlos Berenstein on his 60*th birthday.*

Summary. We study a class of localized solutions of the wave equation, called *eigenwavelets*, obtained by extending its fundamental solutions to complex spacetime in the sense of hyperfunctions. The imaginary spacetime variables y, which form a timelike vector, act as *scale parameters* generalizing the scale variable of wavelets in one dimension. They determine the *shape* of the wavelets in spacetime, making them *pulsed beams* that can be focused as tightly as desired around a single ray by letting y approach the light cone. Furthermore, the absence of any sidelobes makes them especially attractive for communications, remote sensing and other applications using acoustic waves. (A similar set of "electromagnetic eigenwavelets" exists for Maxwell's equations.) I review the basic ideas in Minkowski space $\mathbb{R}^{3,1}$, then compute sources whose realization should make it possible to radiate and absorb such wavelets. This motivates an extension of Huygens' principle allowing equivalent sources to be represented on shells instead of surfaces surrounding a bounded source.

1 Extension of wave functions to complex spacetime

The ideas to be presented here affirm that complex analysis resonates deeply in "real" physical and geometric settings, and so they are close in spirit to the work of Carlos Berenstein (see [BG91, BG95, B98], for example), to whom this volume is dedicated.

Acoustic and electromagnetic wavelets were first constructed in [K94]. It was shown that solutions of *homogeneous* (i.e., sourceless) scalar and vector wave equations in Minkowski space $\mathbb{R}^{3,1}$ extend naturally to complex spacetime, and the wavelets were defined as the Riesz duals of *evaluation maps* acting on spaces of such holomorphic solutions. The sourceless wavelets then split naturally into retarded and advanced parts emitted and absorbed, respectively, by sources located on *branch cuts* needed to make these parts single valued. Later work [K3, K4] was aimed at the construction of *realizable* source distributions which, when synthesized, would act as antennas radiating and receiving the wavelets. Two difficulties with this approach

have been (a) that the computed sources are quite singular, consisting of multiple surface layers that may be difficult to realize in practice, and (b) in the electromagnetic
case the sources appeared to require a nonvanishing magnetic charge distribution,
which cannot be realized as no magnetic monopoles have been observed in nature.
In this paper, we resolve the first difficulty by replacing the spheroidal surface sup-
porting the sources in [K3, K4] by a spheroidal *shell*. It is shown in [K4a] that the
second difficulty can be overcome using Hertz potentials, which give a charge-current
distribution due solely to *bound electric charges* confined to the shell.

Although our constructions generalize to other dimensions, we shall concentrate
here on the physical case of the Minkowski space $\mathbb{R}^{3,1}$. Let

$$x = (\boldsymbol{r}, t), \qquad y = (\boldsymbol{a}, b) \in \mathbb{R}^{3,1} \tag{1}$$

be real spacetime vectors and define the complex *causal tube*

$$\mathcal{T} = \{x - iy \in \mathbb{C}^4 : y \text{ is timelike, i.e., } |b| > |\boldsymbol{a}|\}. \tag{2}$$

It was shown in [K94, K3] that solutions of the *homogeneous* wave equation

$$\Box f_0(x) \equiv (\partial_t^2 - \Delta) f_0(\boldsymbol{r}, t) = 0 \tag{3}$$

extend naturally to analytic functions $\tilde{f}_0(x - iy)$ in $\mathcal{T}$ in the sense that

$$\lim_{y \to +0} \{\tilde{f}_0(x - iy) - \tilde{f}_0(x + iy)\} = f_0(x), \tag{4}$$

where $y \to +0$ means that y approaches the origin within the future cone, i.e.,
with $b > |\boldsymbol{a}|$. This kind of extension to complex domains is familiar in hyperfunction
theory; see [K88, KS99], for example. We now show that even when the wave function
has a source, i.e.,

$$\Box f(x) = 4\pi g(x), \tag{5}$$

it extends analytically to $\mathcal{T}$ outside a spacetime region determined by the source. It
will suffice to do this for the *retarded propagator*

$$G(x) = \frac{\delta(t - r)}{r}, \tag{6}$$

which is the unique causal fundamental solution:

$$\Box G(x) = 4\pi \delta(t)\delta(\boldsymbol{r}) = 4\pi \delta(x), \quad G(\boldsymbol{r}, t) = 0 \quad \forall t < 0. \tag{7}$$

If the source g is supported in a compact spacetime region W, the unique causal
solution of (5) is given by

$$f(x) = \int_W dx' G(x - x')g(x'). \tag{8}$$

Assume for the moment that $G(x)$ has been extended to $\tilde{G}(x - iy)$. Then we *define*
the source of $\tilde{G}$ as the distribution $\tilde{\delta}$ in *real* spacetime given by

$$4\pi \tilde{\delta}(x - iy) \equiv \Box_x \tilde{G}(x - iy), \tag{9}$$

where $\Box_x$ means that the wave operator acts only on x, in a distributional sense, so that the imaginary spacetime vector y is regarded as an auxiliary parameter. The extended solution is now defined as

$$\tilde{f}(x - iy) = \int_W dx' \tilde{G}(x - x' - iy)g(x') \tag{10}$$

and it satisfies the wave equation

$$\Box_x \tilde{f}(x - iy) = 4\pi \tilde{g}(x - iy)$$

with the extended source

$$\tilde{g}(x - iy) = \int_W dx' \tilde{\delta}(x - x' - iy)g(x'). \tag{11}$$

Formally, the extended delta function $\tilde{\delta}(x - iy)$ is a "point source" at the imaginary spacetime point iy as seen by a real observer at x. Actually, it will be seen to be a distribution in x with compact *spatial* (but not temporal) support localized around the spatial origin ($r = 0$) and depending on the choice of a *branch cut* needed to make $\tilde{G}$ single valued. This branch cut is precisely the region where $\tilde{G}$ fails to be analytic, and the integral (10) determines a region $\tilde{W}$ containing W where $\tilde{f}$ fails to be analytic.

A *general* solution $f_1(x)$ of (5) is obtained by adding a sourceless wave $f_0(x)$ to (8). Since $\tilde{f}_0$ is analytic in $\mathcal{T}$, $\tilde{f}_1$ is analytic in $\mathcal{T}$ outside of $\tilde{W}$. It therefore suffices to concentrate on the propagators as claimed. In the rest of the paper, we construct extended propagators, study their properties, and compute their sources.

2 Extended propagators

In accordance with (1), we use the following notation for complex space and time variables:

$$\tilde{\boldsymbol{r}} = \boldsymbol{r} - i\boldsymbol{a} \in \mathbb{C}^3, \qquad \tilde{t} = t - ib \in \mathbb{C},$$
$$\tilde{x} = x - iy = (\tilde{\boldsymbol{r}}, \tilde{t}) \in \mathcal{T} \Leftrightarrow |b| > |\boldsymbol{a}|.$$

As above, we interpret $i\boldsymbol{a}$ formally as an imaginary spatial source point so that $\tilde{\boldsymbol{r}}$ is the vector from the imaginary source point $i\boldsymbol{a}$ to a real observer at $\boldsymbol{r}$. To extend the propagator (6), begin by replacing the one-dimensional delta function with the *Cauchy kernel*,

$$\delta(t) \to \tilde{\delta}(\tilde{t}) = \frac{1}{2\pi i \tilde{t}}, \qquad \tilde{t} = t - ib, \tag{12}$$

which indeed satisfies a condition of type (4):

$$\lim_{b \to +0} \{\tilde{\delta}(t - ib) - \tilde{\delta}(t + ib)\} = \delta(t). \tag{13}$$

To complete the extension of $G(r, t)$, we must also extend the Euclidean distance $r(r) = |r|$. Define the *complex distance* from the source to the observer as

$$\tilde{r}(\tilde{r}) = \sqrt{\tilde{r} \cdot \tilde{r}} = \sqrt{r^2 - a^2 - 2ir \cdot a}, \quad \text{where } r = |r|, \quad a = |a|. \quad (14)$$

$\tilde{r}(\tilde{r})$ is an analytic continuation to $\mathbb{C}^3$ of $r(r)$. Being a complex square root, it has branch points wherever $\tilde{r} \cdot \tilde{r} = 0$. For fixed $a \neq 0$, these form a circle of radius a in the plane orthogonal to a,[1]

$$C \equiv \{r \in \mathbb{R}^3 : \tilde{r} = 0\} = \{r : r = a, \ r \cdot a = 0\}. \quad (15)$$

To be consistent with the notation $\tilde{r} = r - ia$, we write

$$\tilde{r} = p - iq. \quad (16)$$

Comparison with (14) gives the following relations between (p, q) and the spherical and cylindrical coordinates with axis along a:

$$p^2 - q^2 = r^2 - a^2, \quad pq = a \cdot r = ar \cos \theta = az \quad (17)$$

and

$$\begin{aligned} a^2 \rho^2 &= a^2(r^2 - z^2) = a^2(a^2 + p^2 - q^2) - p^2 q^2 \\ &= (a^2 + p^2)(a^2 - q^2). \end{aligned} \quad (18)$$

It follows that the real and imaginary parts of $\tilde{r}$ are bounded by r and a, respectively:

$$\begin{aligned} p^2 &\leq r^2, \quad \text{i.e.,} \quad |\operatorname{Re} \tilde{r}| \leq |\operatorname{Re} \tilde{r}|, \\ q^2 &\leq a^2, \quad \text{i.e.,} \quad |\operatorname{Im} \tilde{r}| \leq |\operatorname{Im} \tilde{r}|, \end{aligned} \quad (19)$$

with equalities attained only when r is parallel or antiparallel to a.

Since a will be a fixed nonzero vector throughout, we will usually regard $\tilde{r}, p, q$ as functions of r only, suppressing the dependence on a. Note that $\mathbb{R}^3 - C$ is multiply connected since a closed loop that threads C cannot be shrunk continuously to a point without intersecting C. In particular, if we continue $\tilde{r}$ analytically around a simple closed loop, we obtain the value $-\tilde{r}$ instead of $\tilde{r}$ upon returning to the starting point. Thus $\tilde{r}$ is a double-valued function on $\mathbb{R}^3$. To make it single valued, we choose a branch cut that must be crossed to close the loop. Instead of returning to the starting point as $-\tilde{r}$, the sign reversal now takes place upon crossing the cut. To give an extension of the *positive* distance, the branch must be chosen so that

$$a \to 0 \Rightarrow \tilde{r} \to +r, \quad (20)$$

and the simplest such choice is obtained by requiring

$$\operatorname{Re} \tilde{r} = p \geq 0. \quad (21)$$

---

[1] In $\mathbb{R}^n$, C would be a sphere of codimension 2 orthogonal to a.

The resulting branch cut consists of the disk spanning the circle C,

$$\mathcal{D} \equiv \{r \in \mathbb{R}^3 : p = 0\} = \{r : r \leq a,\ r \cdot a = 0\}, \quad \partial \mathcal{D} = C. \tag{22}$$

$\mathcal{D}$ will be called the *standard branch cut* and $\tilde{r}$ the *standard complex distance*. General branch cuts, obtained by deforming $\mathcal{D}$ while leaving its boundary intact, will be considered in the next section.

If the observer is far from C, it follows from (14) and (20) that

$$r \gg a \Rightarrow p \approx r \quad \text{and} \quad q \approx a \cos \theta. \tag{23}$$

Thus $(p, q/a)$ are deformations of the spherical coordinates $(r, \cos \theta)$ near the source. From (17) and (18), it follows that level surfaces of p^2 (as a function of r, keeping $a \neq 0$ fixed) are spheroids $\mathcal{S}_p$ and those of q^2 are the orthogonal hyperboloids $\mathcal{H}_q$, given by

$$\mathcal{S}_p : \frac{\rho^2}{p^2 + a^2} + \frac{z^2}{p^2} = 1, \quad p \neq 0, \tag{24}$$

$$\mathcal{H}_q : \frac{\rho^2}{q^2 - a^2} - \frac{z^2}{q^2} = 1, \quad 0 < q^2 < a^2. \tag{25}$$

All these quadrics are *confocal* with C as the common focal set. As $p \to 0$, $\mathcal{S}_p$ collapses to a double cover of the disk $\mathcal{D}$. The variables (p, q), together with the azimuthal angle ϕ about the a-axis, determine an *oblate spheroidal coordinate system*, as depicted in Figure 1.

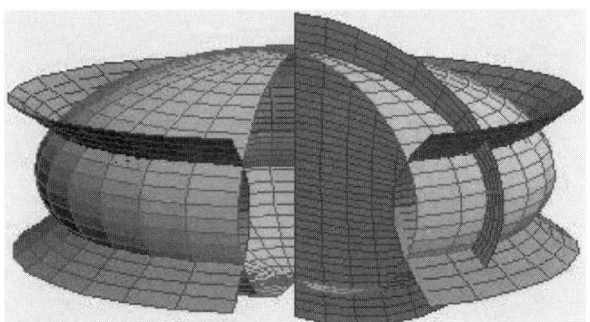

Fig. 1. The level surfaces of p, q, and ϕ form an oblate spheroidal coordinate system.

We now define the extended propagator as

$$\tilde{G}(\tilde{r}, \tilde{t}) = \frac{\tilde{\delta}(\tilde{t} - \tilde{r})}{\tilde{r}} = \frac{1}{2\pi i \tilde{r}(\tilde{t} - \tilde{r})}. \tag{26}$$

This is our *basic wavelet*,[2] from which the entire wavelet family is obtained by spacetime translations:

---

[2] In applications, it is better to use time derivatives of $\tilde{G}$, which have vanishing moments and better temporal decay and propagation properties [K4].

$$\tilde{G}_z(x) = \tilde{G}(x - z), \quad z = x' + iy = (\boldsymbol{r}' + i\boldsymbol{a}, t' + ib) \in \mathcal{T}. \qquad (27)$$

The family $\tilde{G}_z$ may be called *eigenwavelets* of the wave equation in the sense that they are *proper* to that equation, though of course they are not eigen*functions*. In fact, $\tilde{G}_z(x)$ is seen [K3] to be a *pulsed beam* originating from $\boldsymbol{r} = \boldsymbol{r}'$ at $t = t'$ and propagating along the direction of $\boldsymbol{a}/b$, i.e., along $\boldsymbol{a}$ if y is in the future cone and along $-\boldsymbol{a}$ if y is in the past cone. The pulse has a duration $|b| - a$ along the beam axis. By letting y approach the light cone ($a \to |b|$), the beam can be focused as tightly as desired around its axis, approximating a single *ray* along y. Equation (10) states that the extended causal solution $\tilde{f}(x - iy)$ is a superposition of eigenwavelets, all with the same y. This gives a *directional scale analysis* of the original solution $f(x)$ which may be called its *eigenwavelet transform*.

The eigenwavelets have the spheroids $\mathcal{S}_p$ as *wave fronts* and propagate out along the orthogonal hyperboloids $\mathcal{H}_q$ with strength decaying monotonically away from the front beam axis. Hence they have no *sidelobes*, which makes them potentially useful for applications to communication, radar and related areas. These properties are illustrated in Figures 2 and 3.

We may visualize the effects of the extension $G(\boldsymbol{r}, t) \to \tilde{G}(\tilde{\boldsymbol{r}}, \tilde{t})$ as follows. The extension $t \to \tilde{t}$ replaces the spherical *impulse* $\delta(t - r)$ in (6) by a *spherical pulse*

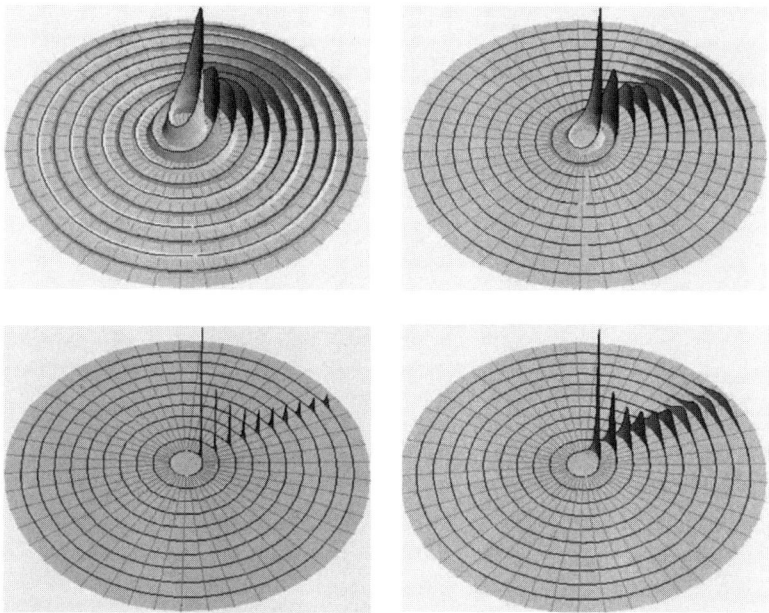

Fig. 2. Time-lapse plots of $|\tilde{G}(x - iy)|$ in the far zone, showing the evolution of a *single pulse* with propagation vector $y = (0, 0, a, b)$. *Clockwise from upper left:* $b/a = 1.5$, 1.1, 1.01, 1.0001. As $b/a \to 1$, y approaches the light cone and the pulsed beam becomes more and more focused around the ray y. We have taken the slice $x_2 = 0$ so that the disk $\mathcal{D}$ becomes the interval $[-a, a]$ on the x_1-axis and the pulse propagates in the x_3 direction of the x_1–x_3 plane.

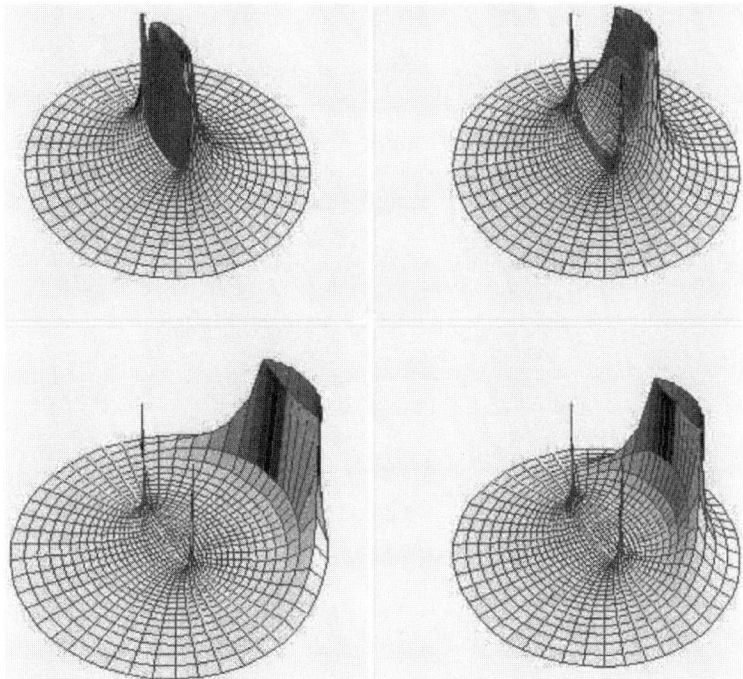

Fig. 3. Near-zone graphs of $|\tilde{G}(x - iy)|^2$ with $y = (0, 0, 1, 1.01)$ immediately after launch, evolving in the x_1–x_3 plane with $x_2 = 0$ as in Figure 1. *Clockwise from upper left: $t =$* 0.1, 1, 2, 3. The ellipsoidal wave fronts and hyperbolic flow lines are visible. The top of the peak is cut off to show the behavior near the base. The two spikes represent the branch circle, whose slice with $x_2 = 0$ consists of the points $(\pm 1, 0, 0)$.

$\tilde{\delta}(\tilde{t} - r)$ of duration $|b|$. The extension $r \to \tilde{r}$ then *deforms* this spherical pulse to a pulsed beam in the direction of a/b. By (23),

$$r \gg a \Rightarrow \tilde{r} \approx r - ia\cos\theta; \tag{28}$$

hence the larger we choose a, the stronger the dependence of $\tilde{r}$ on $\cos\theta$ in the far zone and the more *focused* the beam.

Let us emphasize that $\tilde{G}$ depends on the complex spatial vector $\tilde{\boldsymbol{r}} \in \mathbb{C}^3$ only through the complex distance $\tilde{r}$ by writing

$$\Psi(\tilde{r}, \tilde{t}) = \tilde{G}(\tilde{r}, \tilde{t}) = \frac{1}{2\pi i \tilde{r}(\tilde{t} - \tilde{r})}. \tag{29}$$

Due to the factor $\tilde{r}$ in the denominator, Ψ is discontinuous across $\mathcal{D}$ and singular on $\mathcal{C}$. $\mathcal{D}$ generalizes the point singularity of G at $\boldsymbol{r} = \boldsymbol{0}$ and will be the spatial support of the source (9). To avoid further singularities, the factor

$$\tilde{t} - \tilde{r} = (t - p) - i(b - q)$$

must not vanish for any r. By (19),

$$b - q \neq 0 \quad \forall r \Leftrightarrow a < |b|, \tag{30}$$

so a necessary and sufficient condition for $\Psi(\tilde{r}, \tilde{t})$ to be analytic whenever $r \notin \mathcal{D}$ is that $(\tilde{r}, \tilde{t}) \in \mathcal{T}$. Recalling that the tightness of the beam is controlled by the size of a, (30) means that the beam cannot become tighter than a single ray and, in fact, fails to be analytic *along the ray* in the limit $a = |b|$.

The volume element in $\mathbb{R}^3$ in oblate spheroidal coordinates is

$$dV = \frac{1}{a}(p^2 + q^2)dpdqd\phi = \frac{1}{a}|\tilde{r}|^2 dpdqd\phi; \tag{31}$$

hence Ψ is *locally* integrable and square integrable. A differentiation gives

$$4\pi \tilde{\delta}(x - iy) \equiv \Box_x \tilde{G}(x - iy) = 0, \quad (x - iy \in \mathcal{T}, \; r \notin \mathcal{D}). \tag{32}$$

Therefore, $\tilde{\delta}(x - iy)$, with y a fixed timelike vector, is a distribution in $x = (r, t)$ with *spatial* support in $\mathcal{D}$. (The *temporal* support is noncompact; in fact, $\tilde{\delta}(x - iy)$ decays as $1/\tilde{t}$ due to the Cauchy kernel.)

The source $\tilde{\delta}(x - iy)$ was computed explicitly in [K3] and turns out to be quite singular. It consists of a single layer and a double layer on $\mathcal{D}$, both of which diverge on the boundary $\mathcal{C}$ where Ψ is singular. We will compute *regularized* versions of Ψ and $\tilde{\delta}$ by using the freedom to deform the branch cut to eliminate the singularity on $\mathcal{C}$.

3 Regularization by branch cut deformation

A general branch cut $\mathcal{B}$ is a *membrane* obtained by a continuous deformation of the disk $\mathcal{D}$ leaving its boundary intact,

$$\partial \mathcal{B} = \mathcal{C}. \tag{33}$$

$\mathcal{B}$ inherits an orientation from $\mathcal{D}$, which in turn is oriented by a. Let $V_\mathcal{B}$ be the compact volume swept out in the deformation from $\mathcal{D}$ to $\mathcal{B}$. Let us define the complex distance $\tilde{r}_\mathcal{B}$ with branch cut $\mathcal{B}$ in terms of $\tilde{r} = \tilde{r}_\mathcal{D}$ by

$$\tilde{r}_\mathcal{B} = \begin{cases} \tilde{r} & \text{if } r \notin V_\mathcal{B}, \\ -\tilde{r} & \text{if } r \in V_\mathcal{B}. \end{cases} \tag{34}$$

I claim that $\tilde{r}_\mathcal{B}$ *is continuous except for a sign reversal across* $\mathcal{B}$, generalizing the sign reversal of $\tilde{r}$ across $\mathcal{D}$. This can be seen most simply if $\mathcal{B}$ does not intersect the interior of $\mathcal{D}$, so that they have only the boundary in common. Then $V_\mathcal{B}$ is either all on the "positive" or all on the "negative" side of $\mathcal{D}$. If $V_\mathcal{B}$ is "positive," then its boundary is

$$\partial V_\mathcal{B} = \mathcal{B} - \mathcal{D}, \tag{35}$$

meaning that the orientation (outward normal) of the boundary is positive on $\mathcal{B}$ and negative on $\mathcal{D}$. Since $\tilde{r}$ changes sign upon crossing $\mathcal{D}$ "upward" into $V_{\mathcal{B}}$, (34) shows that $\tilde{r}_{\mathcal{B}}$ is continuous across $\mathcal{D}$. This proves that its only discontinuity is the sign reversal in crossing $\mathcal{B}$, as claimed. Similarly, if $V_{\mathcal{B}}$ is "negative," then its boundary is

$$\partial V_{\mathcal{B}} = \mathcal{D} - \mathcal{B} \tag{36}$$

and the above argument remains valid. To handle branch cuts that intersect the interior of $\mathcal{D}$, we restate the "negative" case (36) by declaring $V_{\mathcal{B}}$ *negatively oriented*, so that its boundary is oriented by the *inward normal*. Denoting the negatively oriented volume by $-V_{\mathcal{B}}$, (36) can be restated as

$$\partial(-V_{\mathcal{B}}) = \mathcal{B} - \mathcal{D}. \tag{37}$$

Hence the rule (35) applies to every branch cut $\mathcal{B}$ obtained by a continuous deformation of $\mathcal{D}$, whether or not it intersects the interior of $\mathcal{D}$, provided the orientation of the swept-out volume is taken into account. If $\mathcal{B}$ intersects the interior of $\mathcal{D}$, then $V_{\mathcal{B}}$ has both positively and negatively oriented components. The definition (34) of the branch $\tilde{r}_{\mathcal{B}}$ remains valid whether $V_{\mathcal{B}}$ (or any of its components) is positively or negatively oriented.

Of special interest will be the *upper and lower spheroidal branch cuts*

$$\mathcal{B}_{\alpha}^{\pm} = \mathcal{S}_{\alpha}^{\pm} \cup \mathcal{A}_{\alpha}, \tag{38}$$

where $\mathcal{S}_{\alpha}^{\pm}$ denote the upper and lower hemispheroids

$$\mathcal{S}_{\alpha}^{\pm} = \{r \in \mathcal{S}_{\alpha} : \pm z > 0\}$$

and

$$\mathcal{A}_{\alpha} = \{r : r \cdot a = 0, \ a^2 \le r^2 \le \alpha^2 + a^2\}$$

is the *apron* connecting them to $\mathcal{C}$, which must be included so that $\partial \mathcal{B}_{\alpha}^{\pm} = \mathcal{C}$, as required. The cut $\mathcal{B}_{\alpha}^{+}$ is depicted in Figure 4.

We can now construct a *regularized* version of the extended propagator Ψ by taking the *average* of the propagators with cuts $\mathcal{B}_{\alpha}^{+}$ and $\mathcal{B}_{\alpha}^{-}$. Denote the complex distances with cuts $\mathcal{B}_{\alpha}^{\pm}$ by $\tilde{r}_{\pm}$ instead of $\tilde{r}_{\mathcal{B}_{\alpha}^{\pm}}$, and let

$$\Psi_A(\tilde{r}, \tilde{t}) = \frac{1}{2}\{\Psi(\tilde{r}_{+}, \tilde{t}) + \Psi(\tilde{r}_{-}, \tilde{t})\} = \tilde{G}_A(x - iy). \tag{39}$$

Let $V_{\alpha}^{\pm}$ be the interiors of the upper and lower hemispheroids. By (34),

$$r \in V_{\alpha}^{+} \Rightarrow \tilde{r}_{+} = -\tilde{r}, \quad \tilde{r}_{-} = \tilde{r}, \tag{40}$$

$$r \in V_{\alpha}^{-} \Rightarrow \tilde{r}_{+} = \tilde{r}, \quad \tilde{r}_{-} = -\tilde{r}. \tag{41}$$

Hence in both $V_{\alpha}^{\pm}$, we have

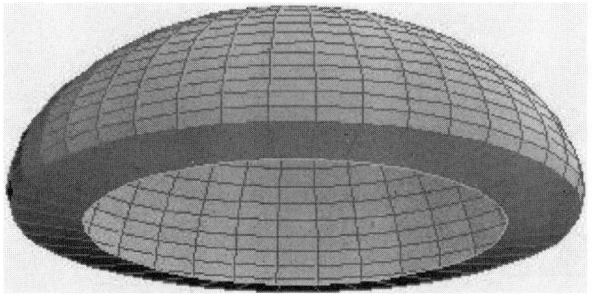

Fig. 4. The upper hemispheroidal branch cut $\mathcal{B}_\alpha^+$ with its apron.

$$\Psi_A(\tilde{r}, \tilde{t}) = \frac{1}{4\pi i \tilde{r}(\tilde{t} - \tilde{r})} - \frac{1}{4\pi i \tilde{r}(\tilde{t} + \tilde{r})} = \frac{1}{2\pi i (\tilde{t}^2 - \tilde{r}^2)}, \tag{42}$$

which is *independent* of the choice of branch. This shows that the discontinuities across the aprons cancel in the average Ψ_A. Furthermore, by (30) we have

$$|b| > a \Rightarrow \tilde{t}^2 - \tilde{r}^2 = (\tilde{t} - \tilde{r})(\tilde{t} + \tilde{r}) \neq 0,$$

showing that the singularities on $\mathcal{C}$ cancel as well. That is, Ψ_A *is analytic at all interior points of the spheroid* $\mathcal{S}_\alpha$.

In the *exterior* of $\mathcal{S}_\alpha$, we have $\tilde{r}_\pm = \tilde{r}$ and hence $\Psi_A = \Psi$. Since $\mathcal{D}$ is contained in $\mathcal{S}_\alpha$ and Ψ is analytic outside of $\mathcal{D}$, we conclude that $\Psi_A(\tilde{r}, \tilde{t})$ fails to be analytic only when $r \in \mathcal{S}_\alpha$. Denoting the interior field by Ψ_1 and the exterior field by Ψ_2, we have

$$\Psi_1(\tilde{r}, \tilde{t}) = \frac{1}{2}\{\Psi(\tilde{r}, \tilde{t}) + \Psi(-\tilde{r}, \tilde{t})\}, \tag{43}$$

$$\Psi_2(\tilde{r}, \tilde{t}) = \Psi(\tilde{r}, \tilde{t}).$$

Thus Ψ_A is analytic except for a *bounded jump discontinuity* across $\mathcal{S}_\alpha$ given by

$$\Psi_J(\tilde{r}, \tilde{t}) \equiv \Psi_2(\tilde{r}, \tilde{t}) - \Psi_1(\tilde{r}, \tilde{t}) = \frac{1}{2}\{\Psi(\tilde{r}, \tilde{t}) - \Psi(-\tilde{r}, \tilde{t})\} = \tilde{G}_J(x - iy). \tag{44}$$

It follows by the same argument as in (32) that the source distribution

$$4\pi \tilde{\delta}_A(x - iy) \equiv \Box_x \Psi_A(\tilde{r}, \tilde{t}) \tag{45}$$

is supported spatially on $\mathcal{S}_\alpha$. Because $\tilde{\delta}_A$ is obtained by twice differentiating a discontinuous function, it consists of a combination of single and double layers on $\mathcal{S}_\alpha$. But the jump discontinuity in Ψ_A is *bounded* (unlike that in Ψ, which diverges on $\mathcal{C}$), and so are these layers; see [K4].

The above arguments remain valid if instead of $\mathcal{B}_\alpha^\pm$ we use *any* two branch cuts whose common interior V contains the branch circle $\mathcal{C}$. In that case, the averaged propagator is analytic in $\mathcal{T}$ except for a finite discontinuity when r crosses the boundary ∂V, and its source distribution is supported spatially on ∂V. However, the above choice has the advantage that $\partial V = \mathcal{S}_\alpha$ are *wave fronts*, hence all parts of the surface radiate simultaneously and coherently.

4 Extended Huygens sources

Let H be the Heaviside step function. Since $0 \leq p < \alpha$ in the interior of S_α and $p > \alpha$ in the exterior, we have

$$\Psi_A(\tilde{r}, \tilde{t}) = H(\alpha - p)\Psi_1(\tilde{r}, \tilde{t}) + H(p - \alpha)\Psi_2(\tilde{r}, \tilde{t}), \qquad (46)$$

where the interior and exterior fields are given by (43). This can be used to compute the source distribution $\tilde{\delta}_A$ defined in (45), and the result is the sum of terms with factors $\delta(p - \alpha)$ and $\delta'(p - \alpha)$. The former are interpreted as *single layers* on S_α, and the latter as *double layers*.

An interesting practical question is whether the wavelets Ψ_A, interpreted as acoustic pulsed beams, can be *realized* by manufacturing their sources. A similar question can be posed for their electromagnetic counterparts, which solve Maxwell's equations; see [K4]. It is doubtful whether an acoustic source can be produced including double layers, and the problem becomes even more difficult in the electromagnetic case because the current density involves yet another derivative, hence a still higher layer [K4a]. The multilayered structure is unavoidable as long as we insist on *surface sources*. We now propose a method for constructing solutions of the wave equation where the transition occurs in a *shell* instead of a surface. It will be simpler to present this method initially in a somewhat more general context.

Given a function $p(r, t)$ on $\mathbb{R}^{n,1}$ and two regular values $p_1 < p_2$ in its range, define two time-dependent surfaces and volumes in $\mathbb{R}^n$ by

$$S_1(t) = \{r : p(r, t) = p_1\}, \qquad S_2(t) = \{r : p(r, t) = p_2\},$$
$$V_1(t) = \{r : p(r, t) < p_1\}, \qquad V_2(t) = \{r : p(r, t) > p_2\}.$$

Let f_1, f_2 be solutions of the wave equation in $\mathbb{R}^{n,1}$ with sources g_1, g_2:

$$\Box f_k(r, t) = g_k, \quad k = 1, 2. \qquad (47)$$

We want to construct an *interpolated solution* $f(r, t)$ such that

$$f(r, t) = f_k(r, t) \quad \forall r \in V_k(t) \qquad (48)$$

and compute its source. This can be done by choosing functions $h_k(r, t)$ with

$$h_1(r, t) = \begin{cases} 1, & r \in V_1(t), \\ 0, & r \in V_2(t), \end{cases} \qquad h_2(r, t) = 1 - h_1(r, t) \qquad (49)$$

and letting

$$f = h_1 f_1 + h_2 f_2 \equiv h_k f_k, \qquad (50)$$

where the (Einstein) summation convention is used. The source of f is found to consist of two parts,

$$g = \Box f = g_I + g_T, \qquad (51)$$

where

$$g_I = h_k g_k \tag{52}$$

is an *interpolated source* and

$$g_T = 2\dot{h}_k \dot{f}_k - 2\nabla h_k \cdot \nabla f_k + (\Box h_k) f_k \quad (\dot{f} \equiv \partial_t f) \tag{53}$$

is a *transitional source*, which, by (49), is supported on the transition shell

$$V_T(t) = \{r : p_1 \le p(r, t) \le p_2\} \tag{54}$$

and depends only on the *jump field* $f_J = f_2 - f_1$:

$$g_T = 2\dot{h}_2 \dot{f}_J - 2\nabla h_2 \cdot \nabla f_J + (\Box h_2) f_J. \tag{55}$$

Now suppose that $V_1(t)$ and $V_T(t)$ are compact and we are given only one source g_2, supported in $V_1(t)$. Letting f_2 be its causal field, our objective is to find an *equivalent source* g supported in $V_T(t)$ whose causal field f is identical with f_2 in $V_2(t)$. It suffices to choose *any* solution f_1 whose source g_1 is supported in $V_2(t)$, since the interpolated source (52) then vanishes and hence $g = g_T$. f_1 is a sourceless *internal field* in $V_1(t)$, and the source g_T so constructed on $V_T(t)$ generalizes the idea of a *Huygens source* on a surface surrounding the support of g_2. We may recover the latter by assuming that p is time-independent (hence so are S_k and V_k) and choosing $h_k(r)$ so that

$$\lim_{p_1 \to p_2} \nabla h_2(r) = \delta(p(r) - p_2) n(r),$$

where $n(r)$ is a field of orthogonal vectors on S_2 pointing into V_2. The corresponding scheme in the electromagnetic case reduces to the usual boundary conditions on an interface between two media [K4a].

Returning to $n = 3$ with $p = \operatorname{Re}\tilde{r}$, let $f_k = \Psi_k$ as in (43) and h_k be time-independent (e.g., functions of p only). A *smoothed* version of Ψ_A (39) is

$$\Psi_A^{\mathrm{sm}} = h_1 \Psi_1 + h_2 \Psi_2. \tag{56}$$

Since Ψ_k are sourceless in V_T, (51) gives the smoothed version of (45) as

$$4\pi \tilde{\delta}_A^{\mathrm{sm}} = \Box_x \Psi_A^{\mathrm{sm}} = -2\nabla h_2 \cdot \nabla \Psi_J - (\Delta h_2)\Psi_J, \tag{57}$$

where

$$\Psi_J = \Psi_2 - \Psi_1 = \frac{1}{2}\{\Psi(\tilde{r}, \tilde{t}) - \Psi(-\tilde{r}, \tilde{t})\} = \frac{\tilde{t}}{2\pi i \tilde{r}(\tilde{t}^2 - \tilde{r}^2)}$$

is the jump field from V_1 to V_2 as in (44), but no longer restricted to a single spheroid S_α. If we now let $p_1 \to p_2 = \alpha$ and

$$h_1 = H(\alpha - p), \qquad h_2 = H(p - \alpha),$$

then the transition becomes abrupt on S_α and Ψ_A^{sm} becomes Ψ_A (39). Since

$$\nabla h_2 = \delta(p - \alpha)\nabla p,$$
$$\Delta h_2 = \delta'(p - \alpha)|\nabla p|^2 + \delta(p - \alpha)\Delta p,$$

equation (57) becomes

$$4\pi\tilde{\delta}_A = -2\delta(p - \alpha)\nabla p \cdot \nabla\Psi_J - \delta(p - \alpha)\Delta p\Psi_J - \delta'(p - \alpha)|\nabla p|^2\Psi_J$$

displaying the aforementioned single and double layer structure on $\mathcal{S}_\alpha$. To get an explicit expression, use [K4, Appendix]

$$\nabla p = \frac{pr + qa}{p^2 + q^2}, \qquad \Delta p = \frac{2p}{p^2 + q^2},$$

$$|\nabla p|^2 = \frac{p^2 + a^2}{p^2 + q^2}, \qquad \nabla p \cdot \nabla q = 0,$$

and

$$\nabla p \cdot \nabla\Psi_J = \Psi'_J\nabla p \cdot \nabla\tilde{r} = \Psi'_J|\nabla p|^2,$$

where Ψ'_J is the complex derivative of $\Psi(\tilde{r}, \tilde{t})$ with respect to $\tilde{r}$ (keeping in mind that $\Psi(\pm\tilde{r}, \tilde{t})$ are analytic in $\tilde{r}$ for $p > 0$),

$$\Psi'_J = \frac{\partial\Psi_J}{\partial\tilde{r}} = -\frac{\tilde{t}}{2\pi i\tilde{r}^2(\tilde{t}^2 - \tilde{r}^2)^2}.$$

5 Conclusions

Although I have concentrated on the wave equation in four-dimensional Minkowski space $\mathbb{R}^{3,1}$, similar considerations apply in $\mathbb{R}^{n,1}$. In fact, the awkward extension of the propagator, using the Cauchy kernel in time but the complex distance in space, becomes much more natural when $\tilde{G}(\tilde{r}, \tilde{t})$ is viewed as the retarded part of the analytic continuation of the fundamental solution $G_E(\mathbf{R})$ of Laplace's equation in *Euclidean* $\mathbb{R}^{n+1}$ [K0, K3], based on the complex distance

$$\tilde{R} = \sqrt{\tilde{\mathbf{R}} \cdot \tilde{\mathbf{R}}}, \quad \tilde{\mathbf{R}} \in \mathbb{C}^{n+1},$$

whose branch points form a sphere S^{n-1} in $\mathbb{R}^{n+1}$ of codimension 2 and radius $|\operatorname{Im}\tilde{\mathbf{R}}|$. The extended delta function $\tilde{\delta}_E(\tilde{\mathbf{R}})$,[3] defined by applying the Laplacian in $\mathbf{R}$ to the extension $\tilde{G}_E(\tilde{\mathbf{R}})$, is supported on S^{n-1} for *odd* $n \geq 3$, but a branch cut, consisting of a "membrane" bounded by S^{n-1}, is needed in all other cases.[4] Given any test function f in $\mathbb{R}^{n+1}$, the convolution

$$\tilde{f}(\tilde{\mathbf{R}}) = \int_{\mathbb{R}^{n+1}}\tilde{\delta}_E(\tilde{\mathbf{R}} - \mathbf{R}')f(\mathbf{R}')\,dV(\mathbf{R}') \tag{58}$$

defines an extension of f to $\mathbb{C}^{n+1}$, nonholomorphic in general, whose restriction to the *Minkowski* subspace $\mathbb{R}^{n,1}$, obtained by letting $\tilde{\mathbf{R}} = (\mathbf{r}, it)$, is a solution of the following initial-value problem for the wave equation:

---

[3] The subscript distinguishes $\tilde{\delta}_E(\tilde{\mathbf{R}})$ from the *Minkowskian* $\tilde{\delta}(\tilde{x})$ in (9).
[4] This is because $G_E(\mathbf{R}) = c_n/R^{n-1}$ for $n \geq 2$ and $G_E = c_1 \log R$ for $n = 1$.

$$(\partial_t^2 - \Delta_r)\tilde{f}(r, it) = 0, \tag{59}$$

$$\tilde{f}(r, 0) = f(r, 0), \tag{60}$$

$$(\partial_t - i\partial_b)\tilde{f}(r, b + it) \mid_{b=t=0} = 0. \tag{61}$$

For odd $n \geq 3$, the proof of (59) is based on the fact that $\tilde{\delta}_E$ is distributed uniformly on S^{n-1} and hence $\tilde{f}$ is a *spherical mean* of f [J55]. This relates the support of $\tilde{\delta}_E$ for odd $n \geq 3$ to *Huygens' principle*. The other cases can be treated by applying a distributional version of Hadamard's method of descent.

Equation (61) states that $\tilde{f}(r, b+it)$ satisfies the Cauchy–Riemann equation in its last variable; but since this holds only at one point, it does not imply analyticity—as it cannot since $f(r, b)$ need not have any analytic continuation in b. If one exists, it is indeed given by $\tilde{f}(r, b + it)$. This generalizes an old theorem by Paul Garabedian [G64, pp. 191–202].

Acknowledgments I thank Dr. Arje Nachman for his sustained support of my research, most recently through AFOSR Grant FA9550-04-1-0139.

References

[BG91] C. A. Berenstein and R. Gay, *Complex Variables: An Introduction*, Springer-Verlag, New York, 1991.

[BG95] C. A. Berenstein and R. Gay, *Complex Analysis and Special Topics in Harmonic Analysis*, Springer-Verlag, New York, 1995.

[B98] C. A. Berenstein, *Integral Geometry, Radon Transforms and Complex Analysis: Lectures Given at the 1st Session of the Centro Internazionale Matematico Estivo (C.I.M.E.)*, Lecture Notes in Mathematics 1684, Springer-Verlag, New York, 1998, 1–33.

[G64] P. R. Garabedian, *Partial Differential Equations*, 1st ed., Chelsea, New York, 1964; 2nd revised ed., AMS–Chelsea, Providence, RI, 1998.

[J55] F. John, *Plane Waves and Spherical Means*, Interscience, New York, 1955.

[K88] A. Kaneko, *Introduction to Hyperfunctions*, Kluwer, Dordrecht, the Netherlands, 1988.

[KS99] G. Kato and D. C. Struppa, *Fundamentals of Algebraic Microlocal Analysis*, Marcel Dekker, New York, 1999.

[K94] G. Kaiser, *A Friendly Guide to Wavelets*, Birkhäuser, Boston, 1994; 6th printing, 1999.

[K0] G. Kaiser, Complex-distance potential theory and hyperbolic equations, in J. Ryan and W. Sprössig, eds., *Clifford Analysis*, Birkhäuser, Boston, 2000.

[K3] G. Kaiser, Physical wavelets and their sources: Real physics in complex spacetime, *J. Physics A Math. Gen.*, **36**-30 (2003), R29–R338 (topical review).

[K4] G. Kaiser, Making electromagnetic wavelets, *J. Phys. A Math. Gen.*, **37** (2004), 5929–5947.

[K4a] G. Kaiser, Making electromagnetic wavelets II: Spheroidal shell antennas, preprint, 2004; available online from arxiv.org/abs/math-ph/0408055.

Security Analysis and Extensions of the PCB Algorithm for Distributed Key Generation

Radha Poovendran[1] and Brian Matt[2]

[1] University of Washington
   Seattle, WA 98195-2500
   USA
   radha@ee.washington.edu
[2] McAfee Research
   Rockville, MD 20850-4601
   USA
   brian_matt@mcafee.com

Dedicated to Professor Carlos Berenstein on his 60th birthday.

Summary. Poovendran, Corson, and Baras presented a distributed cryptographic key-generation algorithm that was suitable for wireless networking environment. However, the security as well as the computational complexity of their scheme were never analyzed. In this work, we present information theoretic analysis of their work and derive the properties of the cryptographic keys that are generated by their scheme. We also present efficient computational schemes that would require only a logarithmic number of steps in group size to compute the common keys.

1 Introduction

Broadcast is the inherent mode of communication in wireless networks that deploy omnidirectional antennas. In broadcast mode, all members who are within the communication range of the transmitting node can receive the message, thus making it resource efficient for the sender as well as the network. However, in many applications the set of users that have access to the communication must be restricted. The use of cryptography is one way to restrict the set of members who can access the communication. When the amount of data is high, the use of symmetric keys will help reduce the computational overhead due to the encryption and decryption. However, the use of symmetric keys requires that all members share the same keys for decryption. Several methods have been proposed to generate and distribute a single common key to all the members of a communicating group. Among these methods is the distributed key-generation method proposed by Poovendran, Corson, and Baras in [PCB], which we call the PCB scheme in this paper. The PCB scheme made use of modulo arithmetic and generalized the property of one-time pad, proposed by Shannon [CS]. However, as of now there is no analysis on the security properties of the PCB method. In this work, we enhance the original PCB algorithm and present the security analysis based on information theoretic techniques. We also show how to develop a computationally efficient algorithm for computing the PCB keys.

The organization of the chapter is as follows: We first review the one-time pad and its properties using probabilistic as well as information theoretic approaches. We then present the PCB algorithm. We provide detailed analysis of the PCB algorithm using probabilistic as well as information theoretic techniques. We also show how to develop computationally efficient techniques that will enable efficient calculation of the group's shared key.

2 Properties of the one-time pad–based encryption

We use the notations in [DS] to define a cryptosystem. A cryptographic system is a pentuple $\mathcal{P}, \mathcal{C}, \mathcal{K}, \mathcal{E}, \mathcal{D}$, where the following conditions are satisfied:

1. $\mathcal{P}$ is a finite set of possible messages or *plaintexts*.
2. $\mathcal{C}$ is a finite set of possible encrypted messages or *ciphertexts*.
3. $\mathcal{K}$ is the finite set of keys or the *keyspace*.
4. $\mathcal{E}_K$ is the encryption rule for a given key K. We denote $E_K : \mathcal{P} \to \mathcal{C}$.
5. $\mathcal{D}_K$ is the decryption rule for a given key K. We denote $D_K : \mathcal{C} \to \mathcal{P}$.

2.1 One-time pad cryptosystem

Let p be a large prime. The plaintext and the encryption key are of the same length and chosen independently and are assumed to be picked uniformly in the interval $[0, p-1]$. The encryption rule is the modulo addition with respect to p. The one-time pad scheme is given below:

1. $\mathcal{P} \in \{0, 1, \ldots, p-1\}$.
2. $\mathcal{C} \in \{0, 1, \ldots, p-1\}$.
3. $\mathcal{K} \in \{0, 1, \ldots, p-1\}$.
4. $\mathcal{E}_K$ is a rule $\forall X \in \mathcal{P}, E_K(X) = X + K \bmod p$.
5. $\mathcal{D}_K$ is a rule $\forall Y \in \mathcal{C}, D_K(Y) = Y - K \bmod p$.

If it can be shown that the ciphertext Y is independent of the encryption key or plaintext X, then observing the ciphertext Y reveals no information about the plaintext X, and hence the mutual information [CT] is $I(X \wedge Y) = E_{P_{XY}}[\log \frac{P_{XY}}{P_X P_Y}] = 0$. The main idea behind the one-time pad–based encryption is stated in the following theorem.

Theorem 1. *Let p be a large prime number, and let A and B be two random variables that are mutually independent and uniformly distributed over the interval $[0, p-1]$. Let $C = A + B \bmod p$. Then the random variable C is uniformly distributed in the interval $[0, p-1]$, and the random variables A, B, C are mutually independent.*

Proof. We compute the distribution of C using

$$P(C = k) = \sum_{i=0}^{p-1} P(C = k|a = i)P(A = i) \tag{1}$$

$$= \sum_{i=0}^{p-1} P(A + B = k|A = i)P(A = i) \tag{2}$$

$$= \frac{1}{p} \sum_{i=0}^{p-1} P(B = k - i|A = i) \tag{3}$$

$$= \frac{1}{p} \sum_{i=0}^{p-1} P(B = k - i) \tag{4}$$

$$= \frac{1}{p}. \tag{5}$$

Hence C is uniformly distributed over the range $[0, p - 1]$. We now show that C is independent of A, B:

$$P(C = k|A = i) = P(A + B = k|A = i) \tag{6}$$
$$= P(B = k - i|A = i) \tag{7}$$
$$= P(B = k - i) \tag{8}$$
$$= \frac{1}{p} \tag{9}$$
$$= P(C = k). \tag{10}$$

Hence C is not only uniformly distributed in the interval $[0, p-1]$ but also independent of A (as well as B).

A direct consequence of these derivations is the fact that the random variable C is uncorrelated to random variable A or B. Hence observing random variable C provides no information [CT] about random variables A or B. This idea can be expressed in terms of the mutual information between random variables C and A as

$$I(C \wedge A) = E_{P_{AC}} \left[\log \frac{P_{AC}}{P_A P_C} \right]. \tag{11}$$

Noting that $P_{AC} = P_A P_C$, since A is independent of C, we find that $\log \frac{P_{AC}}{P_A P_C} = \log(1) = 0$. Hence the mutual information between the random variables A and C is zero. Substituting $A = X$, $B = K$, and $C = Y$ in the proof above shows that the one-time pad encryption leads to ciphertext that is uniformly distributed in the interval $[0, p - 1]$ and satisfies $I(Y \wedge X) = 0$ as well as $I(Y \wedge K) = 0$.

3 Review of the PCB scheme

The PCB scheme presented in [PCB] can be viewed as a generalized version of one-time pad encryption. The PCB scheme consists of a trusted third party–based initialization step followed by a distributed key-generation step. We first define relevant notation. Let $E_{K_i}(m)$ denote the encryption of message m with key K_i, and let $A \rightarrow B : m$ denote a message m sent from entity A to entity B. The PCB scheme is described below.

3.1 Initialization

In the initialization step, a trusted third party (TTP) selects n participants of the distributed key-generation scheme labeled $\{M_i\}_{i=1}^{n}$. It is assumed that the TTP shares a pairwise key K_i with member M_i of the group. The TTP chooses a large prime p and generates n uniformly distributed and independent random variables denoted $\alpha_{i,0}$, with $i = 1, \ldots, n$. The TTP computes

$$\theta_0 = \sum_{i=1}^{n} \alpha_{i,0}. \tag{12}$$

The TTP initializes each entity M_i using the message transfer

$$\text{TTP} \rightarrow M_i : E_{K_i}(\alpha_{i,0}, \theta_0). \tag{13}$$

3.2 Broadcast enhanced distributed key generation

The distributed key generation consists of two stages. In the first stage each node generates its contribution and secures and transmits it. In the second stage, each node collects contributions of all other nodes and combines them to generate the group key and its future one-time pad. The original PCB scheme in [PCB] assumed pairwise links between nodes. This procedure is computationally intensive and can be avoided in wireless broadcast environments. We also note that in the original PCB scheme, there was no mechanism to make the participants commit to the shares they would contribute to the group key generation. Lack of commitment makes the original PCB scheme vulnerable to attacks by participants who can bias the final outcome. While we do not elaborate on the key space bias in this work, we eliminate it using a commitment. These two changes are reflected in steps 4 and 5 of the algorithm presented below. At the iteration step j, a participant M_i performs the following operations to generate its share of the distributed key:

1. M_i generates a fractional key $\text{FK}_{i,j}$.
2. M_i generates a hidden fractional key $\text{HFK}_{i,j} = \text{FK}_{i,j} + \alpha_{i,j-1}$.
3. M_i generates a commitment $\text{com}_{i,j} = g^{\text{HFK}_{i,j}}$.
4. $M_i \rightarrow * : \text{com}_{i,j}$.
5. $M_i \rightarrow * : E_{\theta_{j-1}}(\text{HFK}_{i,j})$.

A participant M_i then combines the shares to compute the group key and the fresh one-time pad for its computations. A participant M_i performs the following operations:

1. For all $l \in \{1, \ldots, n\}$, obtain $\text{HFK}_{l,j}$, compute and verify that $g^{\text{HFK}_{l,j}} = \text{com}_{l,j}$. If true, proceed to the next steps; else, terminate.

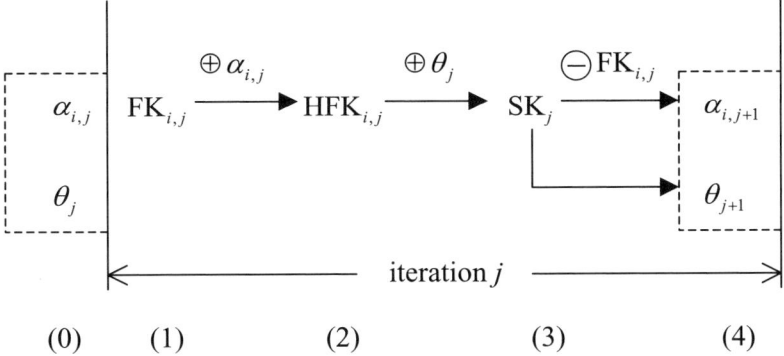

Fig. 1. Iteration and mappings of the key-generation algorithm.

2. Compute the sum of all the hidden fractional keys $\sum_{l=1}^{n} \text{HFK}_{l,j} = \sum_{l=1}^{n} \text{FK}_{l,j} + \sum_{l=1}^{n} \alpha_{l,j-1}$.
3. Compute the new group key as

$$\theta_j = \sum_{l=1}^{n} \text{HFK}_{l,j} + (p-1)\alpha_{l,j-1} = \sum_{l=1}^{n} \text{FK}_{l,j} \bmod p. \qquad (14)$$

4. Compute $\alpha_{i,j} = \theta_j + (p-1)\,\text{FK}_{i,j} \bmod p$.

The PCB scheme is represented in a schematic diagram given in Figure 1.

4 Security analysis of the PCB scheme

As noted earlier, the PCB paper did not provide analysis of the scheme. We provide the security analysis of the PCB scheme in this section. We make the following claims about the security of the PCB scheme.

Theorem 2. *If random variables $\alpha_{i,0}$ are mutually independent and uniformly distributed in the interval $[0, p-1]$, then the group key θ_0, defined by*

$$\theta_0 = \sum_{i=1}^{n} \alpha_{i,0}, \qquad (15)$$

is uniform in the interval $[0, p-1]$ and is mutually independent with respect to any subset consisting of $(n-1)$ of the random variables $\alpha_{i,0}$, $i = 1, \ldots, n$.

Proof. We first show that θ_0 is uniformly distributed and then show that θ_0 is mutually independent of any set of (n-1) $\alpha_{i,0}$. We prove that θ_0 is uniformly distributed using induction. Let $U_i = U_{i-1} + \alpha_{i,0}$ with $U_0 = 0$. Then $U_1 = \alpha_{1,0}, U_2 = \alpha_{1,0} + \alpha_{2,0}, \ldots, U_n = \theta_0$. We now show that $U_i, i = 0, 1, \ldots, n$, are uniformly distributed.

Note that for $i = 1$, $U_1 = \alpha_{1,0}$ is by definition of $\alpha_{1,0}$ uniform over the interval $[0, p - 1]$. For $i = 2$, we have

$$P(U_2 = k) = \sum_{s_1=0}^{p-1} P(U_2 = k|\alpha_{1,0} = s_1)P(\alpha_{1,0} = s_1) \tag{16}$$

$$\overset{(i)}{=} \frac{1}{p} \sum_{s_1=0}^{p-1} P(\alpha_{1,0} = s_1 + \alpha_{2,0} = k|\alpha_{1,0} = s_1) \tag{17}$$

$$= \frac{1}{p} \sum_{s_1=0}^{p-1} P(\alpha_{2,0} = k - s_1|\alpha_1 = s_1) \tag{18}$$

$$\overset{(ii)}{=} \frac{1}{p} \sum_{s_1=0}^{p-1} P(\alpha_{2,0} = k - s_1) \tag{19}$$

$$= \frac{1}{p}. \tag{20}$$

Step (i) follows from the definition of U_2, and step (ii) follows from the observation that under modulo arithmetic the summation includes all the p terms, even if there is an index shift. Hence U_2 is uniformly distributed. Now we show that U_2 is independent of $\alpha_{1,2}$.

$$P(U_2 = k|\alpha_{1,0} = s_1) = P(\alpha_{1,0} = s_1 + \alpha_{2,0} = k|\alpha_{1,0} = s_1) \tag{21}$$

$$= P(\alpha_{2,0} = k - s_1|\alpha_{1,0} = s_1) \tag{22}$$

$$\overset{(i)}{=} P(\alpha_{2,0} = k - s_1) \tag{23}$$

$$= \frac{1}{p} \tag{24}$$

$$= P(U_2 = k). \tag{25}$$

Step (i) follows from the fact that $\alpha_{2,0}$ is independent of $\alpha_{1,0}$. Hence U_2 is independent of U_1; however, $U_2 = \alpha_{2,0} + \alpha_{1,0}$ and $U_1 = \alpha_{1,0}$. Since $\alpha_{1,0}$ and $\alpha_{2,0}$ are mutually independent, interchanging them does not change the result; hence $\sum_{l=1}^{2} \alpha_{l,0}$ is independent of $\alpha_{1,0}$ as well as $\alpha_{2,0}$.

Having illustrated the proof for two variables, let us prove the result for the case that $i = n$ when $\theta_0 = U_n$. We first prove that θ_0 is uniformly distributed and then show it is independent of any subset of $(n - 1)$ αs. For simplicity, we define the notation that $\{Y = y\} = \{\alpha_{i_1,0} = s_{i_1}, \ldots, \alpha_{i_{n-1},0} = s_{i_{n-1}}\}$:

$$P(\theta_0 = k) = P(U_n = k) \tag{26}$$

$$= \sum_{s_1=0}^{p-1} \cdots \sum_{s_{n-1}=0}^{p-1} P(U_n = k|Y = y)P(Y = y)$$

$$= \sum_{s_1=0}^{p-1} \cdots \sum_{s_{n-1}=0}^{p-1} P(Y = y) \prod_{l=1}^{n-1} P(\alpha_{l,0} = s_l) \tag{27}$$

$$= \sum_{s_1=0}^{p-1} \cdots \sum_{s_{n-1}=0}^{p-1} \left\{ \sum_{s_{n-1}=0}^{p-1} P\left(\alpha_{i_n,0} = k - \sum_{i=1}^{n-1} s_i \right) \right\} / p^{n-1} \tag{28}$$

$$= \frac{1}{p}. \tag{29}$$

Hence we note that θ_0 is uniformly distributed in the interval $[0, p-1]$. We now show that θ_0 is independent of any subset of $(n-1)$ αs:

$$P(\theta_0 = k | Y = y) = P(U_n = k | \alpha_{i_1,0} = s_{i_1}, \ldots, \alpha_{i_{n-1},0} = s_{i_{n-1}}) \tag{30}$$

$$= P\left(\sum_{i=1}^{n} \alpha_{i,0} = k | Y = y \right) \tag{31}$$

$$= P\left(\alpha_{i_n,0} = k - \sum_{j=1}^{n-1} s_{i_j} | Y = y \right)$$

$$\overset{(i)}{=} P\left(\alpha_{i_n,0} = k - \sum_{j=1}^{n-1} s_{i_j} \right) \tag{32}$$

$$= \frac{1}{p} \tag{33}$$

$$= P(\theta_0 = k). \tag{34}$$

Step (i) uses the mutual independent property of the αs. Note that the order of picking the αs was random. Hence θ_0 is independent of any arbitrary subset consisting $(n-1)$ αs. We now state the following property of the PCB scheme as a theorem.

Theorem 3. *If random variables $FK_{i,j}$ are mutually independent and uniformly distributed in the interval $[0, p-1]$, then θ_{j+1}, defined by*

$$\theta_j = \sum_{i=1}^{n} FK_{i,j}, \tag{35}$$

is uniform in the interval $[0, p-1]$ and is mutually independent with respect to any subset consisting of $(n-1)$ of the random variables $FK_{i,0}$, $i = 1, \ldots, n$.

Proof. The proof follows the similar inductive argument as above with $\alpha_{i,0}$ replaced with $FK_{i,j}$ and θ_0 replaced with θ_j.

The above theorems show that observing any $(n-1)$ fractional keys does not reveal any information about the group key. Hence an adversary needs to know all n fractional keys to compute the group key θ at any iteration. In terms of the mutual information, we can write

$$I(\theta_j \wedge \text{FK}_{i_1,j}, \ldots, \text{FK}_{i_{n-1},j}) = 0, \tag{36}$$

where the subset of $(n-1)$ fractional keys are chosen arbitrarily.

Theorem 4. *The intermediate pads $\alpha_{i,j}$, computed using the formula*

$$\alpha_{i,j} = \theta_j + (p-1)\,\text{FK}_{i,j} \bmod p, \tag{37}$$

satisfy the property $I(\alpha_{i,j} \wedge \text{FK}_{i,j}) = 0$, $i \in \{1, 2, \ldots, n\}$.

Proof.

$$I(\alpha_{i,j} \wedge \text{FK}_{i,j}) = H(\text{FK}_{i,j}) - H(\text{FK}_{i,j} \,|\, \alpha_{i,j}) \tag{38}$$

$$\stackrel{(i)}{=} H(\text{FK}_{i,j}) - H(\text{FK}_{i,j}) \tag{39}$$

$$= 0. \tag{40}$$

Step (i) follows from the fact that all $\text{FK}_{i,j}$s are mutually independent, and hence $\text{FK}_{l,j}$ is independent of the sum of $\alpha_{i,j} = \sum_{l=1;l \neq i}^{n} \text{FK}_{l,j}$.

Theorem 5. *If the initial parameters $\alpha_{i,0}$s as well as the fractional keys $\text{FK}_{i,j}$s at every computational round j are mutually independent and are uniformly distributed in the interval $[0, p-1]$, then for all j θ_js, the θ_js are uncorrelated.*

Proof. We first show that $I(\theta_j \wedge \theta_m) = 0$ for any arbitrary j, m:

$$I(\theta_j \wedge \theta_m) = H(\theta_j) - H(\theta_j | \theta_m) \tag{41}$$

$$= H\left(\sum_{i=1}^{n} \text{FK}_{i,j}\right) - H\left(\sum_{i=1}^{n} \text{FK}_{i,j} \,\Big|\, \sum_{i=1}^{n} \text{FK}_{i,m}\right) \tag{42}$$

$$\stackrel{(i)}{=} H\left(\sum_{i=1}^{n} \text{FK}_{i,j}\right) - H\left(\sum_{i=1}^{n} \text{FK}_{i,j}\right) \tag{43}$$

$$= 0. \tag{44}$$

Step (i) follows from the fact that given random variables $\text{FK}_{1,j}, \ldots, \text{FK}_{n,j}$ as well as $\text{FK}_{1,m}, \ldots, \text{FK}_{n,m}$ that are mutually independent, any function $f(\text{FK}_{1,j}, \ldots, \text{FK}_{n,j})$ of random variables $\text{FK}_{1,j}, \ldots, \text{FK}_{n,j}$ is independent of any function $g(\text{FK}_{1,m}, \ldots, \text{FK}_{n,m})$ of random variables. For clarity, we use the notation $\{Z\} = \{\sum_{i=1}^{n} \text{FK}_{i,i_1}, \ldots, \sum_{i=1}^{n} \text{FK}_{i,i_m}\}$.

In order to prove the general case, consider the mutual information between a given θ_j and a set $S = \{\theta_{i_1}, \theta_{i_2}, \ldots, \theta_{i_m}\}$, where $\theta_j \in S$. We claim that

$$I(\theta_j \wedge \theta_{i_1}, \theta_{i_2}, \ldots, \theta_{i_m}) = 0. \tag{45}$$

Proof. The proof is similar to the case above but will be presented for completeness:

$$I(\theta_j \wedge \theta_{i_1}, \ldots, \theta_{i_m}) = H(\theta_j) - H(\theta_j | \theta_{i_1}, \ldots, \theta_{i_m}) \tag{46}$$

$$= H\left(\sum_{i=1}^{n} FK_{i,j}\right) - H\left(\sum_{i=1}^{n} FK_{i,j} \mid Z\right) \qquad (47)$$

$$\overset{(i)}{=} H\left(\sum_{i=1}^{n} FK_{i,j}\right) - H\left(\sum_{i=1}^{n} FK_{i,j}\right) \qquad (48)$$

$$= 0. \qquad (49)$$

Again, step (i) follows from the fact that given random variables $FK_{1,j}, \ldots, FK_{n,j}$ as well as $S = \{FK_{1,i_h}, \ldots, FK_{n,i_h}\}_{h=1}^{m}$ that are mutually independent, any function $f(FK_{1,j}, \ldots, FK_{n,j})$ of random variables $FK_{1,j}, \ldots, FK_{n,j}$ is independent of any function $g(FK_{1,i_h}, \ldots, FK_{n,i_m})$ of random variables.

5 Extensions and complexity

Not all wireless networks can be represented by a pure broadcast model. Many networks use multihop communications as well as directional antennas. The impact of directional antennas and multihop communications changes the communication complexity of distributed key generation for some algorithms more than others.

In this section, we describe alternative PCB algorithms better tailored for some nonbroadcast wireless networks. These alternative algorithms are motivated by point-to-point communications in wireless networks. A point-to-point model corresponds to scenarios such as a group of widely distributed members communicating using cell phones, or a localized group communicating using pencil beam directional antennas.

We explore three alternative algorithms for distributed key generation based on hypercube, octopus, and tree structures. We then analyze the communication complexity of the original PCB algorithm, the broadcast-enhanced PCB, and the alternative algorithms. Our analysis has shown that for the point-to-point network, these alternative algorithms have lower communication complexity than either the original or broadcast-enhanced versions of the PCB algorithm. The broadcast-enhanced PCB algorithm has lower complexity than any other algorithms in a pure broadcast network, while the original PCB algorithm has the highest complexity.

Each of the alternative algorithms uses the same initialization phase as the original and broadcast PCB algorithms.

5.1 Hypercube

For simplicity we assume that the group has size $n = 2^r$. Each group member has an identifier i in the range $0, \ldots, n - 1$. In a hypercube, two nodes are connected if their identifiers, represented as binary strings, differ in precisely one position. In the hypercube algorithm, during phase $k = 0, \ldots, r - 1$, each group member communicates with the group member whose identifier differs only in the kth position. After all r phases, each node will have sent and received a message from those group members with which it shares an edge of the hypercube. See Figure 2.

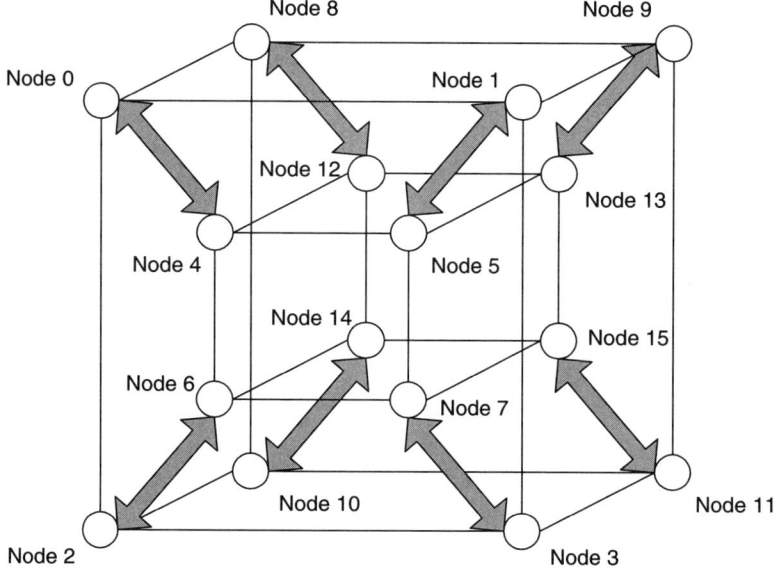

Fig. 2. The hypercube key-generation algorithm with point-to-point communications.

Hypercube algorithm. At the iteration step j, a participant M_i performs the following operations to generate its share of the distributed key:

1. M_i generates a fractional key $FK_{i,j}$.
2. M_i generates a hidden fractional key $HFK_{i,j} = FK_{i,j} + \alpha_{i,j-1}$.
3. For the first set of exchanges in step j, which we call phase $k = 0$,

$$M_i \longrightarrow M_{(\hat{i} = \text{bin}(i) \otimes \text{bin}(2^{k=0}))} : E_{\theta_{j-1}}(KK_{\hat{i},j,0} = HFK_{i,j}),$$

where $\text{bin}(t)$ is the r-bit binary representation of t and $\otimes$ is the exclusive-or operation. Member M_i then computes $TK_{i,j,1} = KK_{i,j,0} + HFK_{i,j}$.
For phases $k = 1, \ldots, r - 1$ of step j,

$$M_i \longrightarrow M_{(\hat{i} = \text{bin}(i) \otimes \text{bin}(2^{k=0}))} : E_{\theta_{j-1}}(KK_{\hat{i},j,k} = TK_{i,j,k-1}).$$

Member M_i then computes $TK_{i,j,k} = KK_{i,j,k-1} + TK_{i,j,k-1}$.
Phase 2 of the hypercube algorithm is shown in Figure 2.

Once the r phases of the exchange are complete, a participant M_i has its combined shares, $\sum_{l=1}^{n} HFK_{l,j} = TK_{i,j,r-1}$. M_i then computes the group key and the fresh one-time pad for its computations. M_i performs the following operations:

1. Compute the new group key as

$$\theta_j = \sum_{l=1}^{n} HFK_{l,j} + (p - 1)\alpha_{l,j-1} = \sum_{l=1}^{n} FK_{l,j} \bmod p.$$

2. Compute $\alpha_{i,j} = \theta_j + (p - 1) FK_{i,j} \bmod p$.

5.2 Octopus-d

The hypercube algorithm provides substantially lower communication complexity than either the original PCB or the broadcast-enhanced PCB algorithms. Further improvement can be achieved by using an octopus network [BW]. An octopus consists of a d-dimension hypercube connecting a core subset of the group with each core member directly connected to a $(2^r - 2^d)/2^d$-size subset of the noncore group members. In Figure 3, a d-dimension hypercube ($d = 2$) is used interconnect 2^d core group members of a group of size $2^r = 16$. Note that if $d = 0$ the octopus network collapses into a star network with a single group member connected to the other $2^r - 1$ members.

In the octopus-d algorithm, each iteration has three passes. During the first pass, each noncore group member transmits its key share to its corresponding core node. In the second pass, the core members perform the exchanges of the hypercube algorithm. During the third pass, each core node passes the sum of the $\mathrm{HFK}_{i,j}$ to its corresponding noncore nodes. See Figure 3.

Octopus-d algorithm. In the algorithm, each core node is numbered according to the hypercube algorithm. Each noncode node is numbered by multiplying the identifier of its corresponding core node by the number of core nodes, 2^d, and adding an index value which runs from 1 to $2^d - 1$.

At the iteration step j, a participant M_i performs the following operations to generate its share of the distributed key:

1. M_i generates a fractional key $\mathrm{FK}_{i,j}$.
2. M_i generates a hidden fractional key $\mathrm{HFK}_{i,j} = \mathrm{FK}_{i,j} + \alpha_{i,j-1}$.
3. Exchanges:
   - *Pass one*: If $2^d \le i \le 2^r - 1$,

$$M_i \longrightarrow M_{\mathrm{core}(i)} : E_{\theta_{j-1}}(\mathrm{HFK}_{i,j}),$$

     where $\mathrm{core}(i)$ is the core group member of i. Pass-one communications are shown in Figure 3. At the end of pass one, each core node computes the sum of its $\mathrm{HFK}_{i,j}$ and those of its dependent noncore group members. Core member i computes

$$\mathrm{KK}_{i,j,0} = \mathrm{HFK}_{i,j} + \sum_{l=i\cdot 2^d+1}^{i\cdot 2^d+2^d-1} \mathrm{HFK}_{l,j} .$$

   - *Pass two*: Use a modified version of the exchanges of the hypercube algorithm on the core group members of the octopus. In the octopus, the values $\mathrm{KK}_{i,j,0}$ are distributed in the first set of exchanges, instead of $\mathrm{HFK}_{i,j}$ used by the standard hypercube algorithm.
   - *Pass three*: If member i is a core member, then depending on the communication model, M_i broadcasts

$$M_i \longrightarrow * : E_{\theta_{j-1}}(\mathrm{TK}_{i,j,d-1})$$

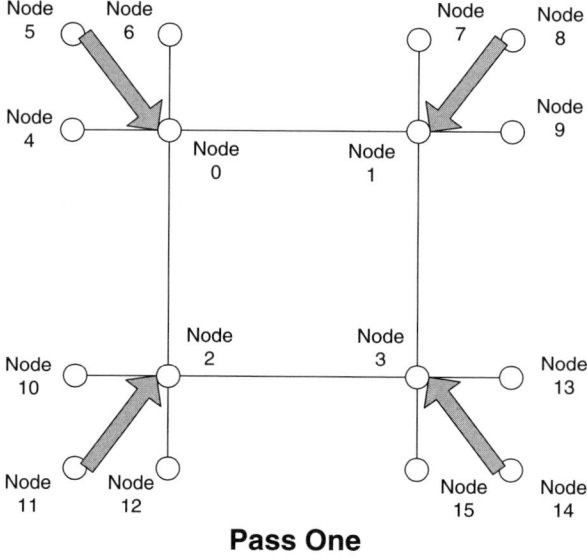

Pass One

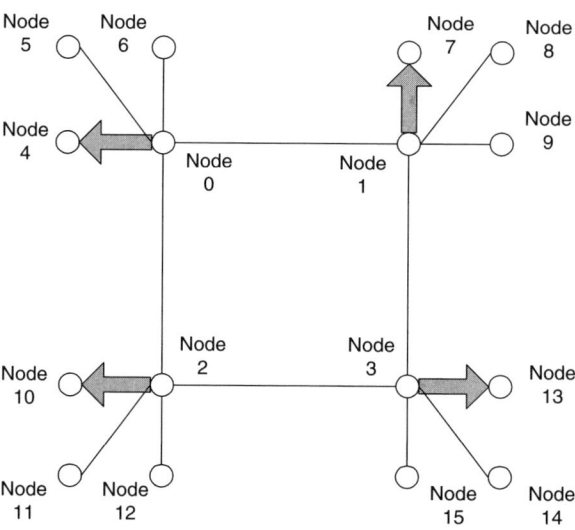

Pass Three

Fig. 3. The octopus-d key-generation algorithm with point-to-point communications ($r = 4$, $d = 2$).

or M_i uses point-to-point messages to exchange $E_{\theta_{j-1}}(\text{TK}_{i,j,d-1})$,

$$M_i \longrightarrow M_{(\text{dependent}_k(i))} : E_{\theta_{j-1}}(\text{TK}_{i,j,d-1}),$$

where $\text{dependent}_k(i)$ is the kth dependent of member i. This phase is shown in Figure 3.

Once the exchanges of this iteration are complete, a participant M_i has its combined shares, $\sum_{l=1}^{n} \text{HFK}_{l,j} = TK_{i,j,r-1}$. M_i then computes the group key and the fresh one-time pad for its computations. M_i performs the following operations:

1. Compute the new group key as

$$\theta_j = \sum_{l=1}^{n} \text{HFK}_{l,j} + (p-1)\alpha_{l,j-1} = \sum_{l=1}^{n} \text{FK}_{l,j} \bmod p.$$

2. Compute $\alpha_{i,j} = \theta_j + (p-1)\,\text{FK}_{i,j} \bmod p$.

5.3 Binary tree

For simplicity, we assume that the group has $n = 2^r - 1$ group members. Each group member has an identifier i in the range $0, \ldots, n-1$ and is a node (interior node or leaf node) of a binary tree. The group members are numbered in order of a preorder tree traversal. Group member 0 is the root of the tree, group member 1 is the left sibling of the root, member 2 is the right sibling of the root, and so on.

Binary tree algorithm. In the tree algorithm, each iteration has two passes. During the first pass, each node of the tree (working from the leaf nodes up toward the root) communicates the sum of the hidden fraction key of all of the node's decedents to its parent. During the second pass, the sum of the hidden fractional keys for the group is distributed by the root. In the point-to-point model, each node of the tree communicates with its children, working from the root down toward the leaf nodes. In the broadcast model, the root distributes the sum directly to each member using a single broadcast. See Figure 4.

At the iteration step j, a participant M_i performs the following operations to generate its share of the distributed key:

1. M_i generates a fractional key $\text{FK}_{i,j}$.
2. M_i generates a hidden fractional key $\text{HFK}_{i,j} = \text{FK}_{i,j} + \alpha_{i,j-1}$.
3. Exchanges:
   - *Pass one*: Pass one propagates hidden fractional keys to the root of the tree. If M_i is a leaf node, i.e., M_I is represented by a level-0 node, then

$$i \longrightarrow \text{parent}(i) : E_{\theta_{j-1}}(\text{HFK}_{i,j}),$$

where $\text{parent}(i)$ is the group member who is represented by the parent node of node i. Group member $M_{\text{parent}(i)}$ then computes $\text{KK}_{\text{parent}(i),j,1} = \text{HFK}_{i,j} + \text{HFK}_{\text{sibling}(i),j} + \text{HFK}_{\text{parent}(i),j}$.

If M_i is represented by a level-k interior node, it must wait until it can compute $\text{KK}_{i,j,k} = \text{KK}_{\text{left\_decendent}(i),j,k-1} + \text{KK}_{\text{right\_decendent}(i),j,k-1} +$

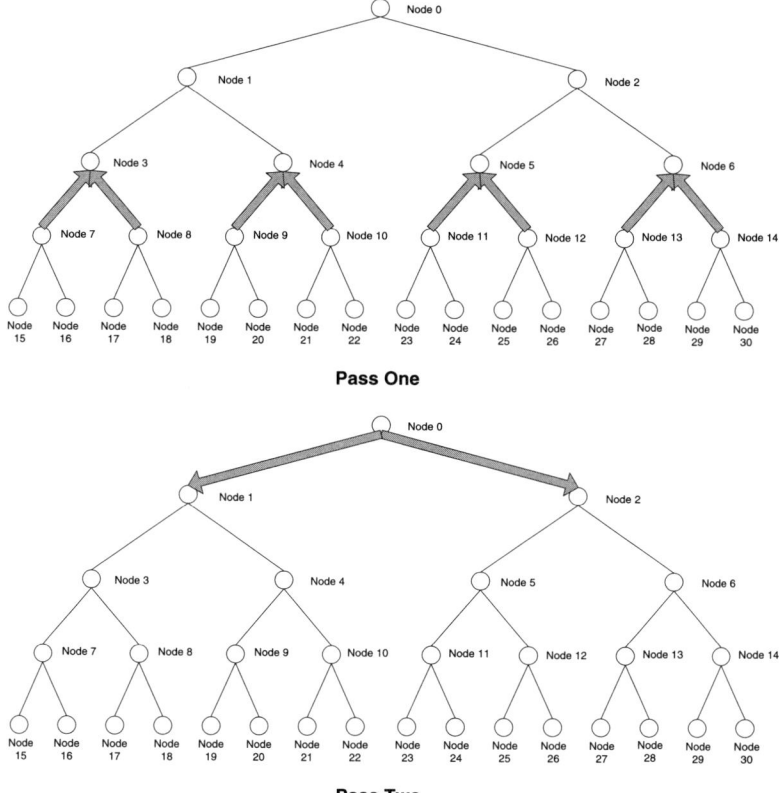

Fig. 4. The tree key-generation algorithm with point-to-point communications.

$HFK_{parent(i),j}$. If M_i is not represented by the root of the tree, then it sends $KK_{i,j,k}$ to its parent, i.e.,

$$i \longrightarrow parent(i) : E_{\theta_{j-1}}(KK_{i,j,k}).$$

Group member $M_{parent(i)}$ then computes $KK_{parent(i),j,k+1} = KK_{i,j,k} + KK_{sibling(i),j,k} + HFK_{parent(i),j}$. Pass one is shown in Figure 4.

• *Pass two*: In the broadcast communication model, the root of the tree broadcasts

$$M_0 \longrightarrow * : E_{\theta_{j-1}}(KK_{0,j,r-1}).$$

In the point-to-point model, the root can distribute $KK_{0,j,r-1}$ to the group using the tree. Each nonleaf, nonroot node M_i receives $KK_{0,j,r-1}$ from its parent and then distributes to its decedents by

$$i \longrightarrow left\_child(i) : E_{\theta_{j-1}}(KK_{i,j,r-1}),$$

$$i \longrightarrow right\_child(i) : E_{\theta_{j-1}}(KK_{i,j,r-1}).$$

Once the exchanges of this iteration are complete, a participant M_i has its combined shares, $\sum_{l=1}^{n} \text{HFK}_{l,j} = \text{KK}_{i,j,r-1}$. M_i then computes the group key and the fresh one-time pad for its computations. M_i performs the following operations:

1. Compute the new group key as

$$\theta_j = \sum_{l=1}^{n} \text{HFK}_{l,j} + (p-1)\alpha_{l,j-1} = \sum_{l=1}^{n} \text{FK}_{l,j} \bmod p.$$

2. Compute $\alpha_{i,j} = \theta_j + (p-1)\text{FK}_{i,j} \bmod p$.

Table 1. Key-generation communication costs: Broadcast.

Algorithm	Group Size	Ave. Trans. per Member	Ave. Recv. per Member	Max. Trans.	Max. Recv.
Orginal PCB	2^r	$2^r - 1$	$(2^r - 1)^2$	$2^r - 1$	$(2^r - 1)^2$
Bcast. PCB	2^r	1	$2^r - 1$	1	$2^r - 1$
Hypercube	2^r	r	$r \cdot (2^r - 1)$	r	$r \cdot (2^r - 1)$
Octopus-d	2^r	$1 + \frac{d}{2^{r-d}}$	$2^r + d \cdot 2^d - 1 - \frac{d-1}{2^{r-d}}$	$d+1$	$2^d \cdot d + 2^r - 1$
Tree	$2^r - 1$	1	$2^r - 2$	1	$2^r - 2$

Table 2. Key-generation communication costs: Point-to-point.

Algorithm	Group Size	Ave. Trans. per Member	Ave. Recv. per Member	Max. Trans.	Max. Recv.
Orginal PCB	2^r	$2^r - 1$	$2^r - 1$	$2^r - 1$	$2^r - 1$
Bcast. PCB	2^r	$2^r - 1$	$2^r - 1$	$2^r - 1$	$2^r - 1$
Hypercube	2^r	r	r	r	r
Octopus-d	2^r	$2 + \frac{d-2}{2^{r-d}}$	$2 + \frac{d-2}{2^{r-d}}$	$2^{r-d} + d - 1$	$2^{r-d} + d - 1$
Tree	$2^r - 1$	$2 + \frac{2}{2^r - 1}$	$2 + \frac{2}{2^r - 1}$	3	3

5.4 Comparison

Tables 1 and 2 compare the communication complexity of the algorithms in the pure broadcast and pure point-to-point models. For each combination of algorithm and model, we give the average number of messages (ignoring commitments) that each group member sends and receives as well as the maximum number of messages sent and received by a group member.

Acknowledgments The authors thank Mingyan Li for useful comments. RP gratefully acknowledges the support of the NSF grants ANI-0093187, ARO grant DAAD-19-02-1-0242, and ONR grant N00014-04-1-0479. Both authors also acknowledge the support of ARL grant DAAD19-01-2-0011.[3]

References

[BW]   K. Becker and I. Wille, Communication complexity of group key distribution, in *Proceedings of the 5th ACM Conference on Computer and Communications Security*, ACM, New York, 1998.

[CT]   T. Cover and J. Thomas, *Elements of Information Theory*, Wiley–Interscience, New York, 1991.

[CS]   C. Shannon, Communication theory of secrecy systems, *Bell Systems Technical J.*, **28** (1949), 656–715.

[DS]   D. Stinson, *Cryptography: Theory and Practice*, CRC Press, New York, 2002.

[PCB] R. Poovendran, S. Corson, and J. Baras, A distributed shared key generation procedure using fractional keys, in *Proceedings of IEEE Milcom*, IEEE, New York, 1998.

---

[3] This document was prepared through collaborative participation in the Communications and Networks Consortium sponsored by the U. S. Army Research Laboratory under the Collaborative Technology Alliance Program, DAAD19-01-2-0011. The U. S. Government is authorized to reproduce and distribute reprints for Government purposes notwithstanding any copyright notation thereon. The views and conclusions contained in this document are those of the author and should not be interpreted as representing the official policies, either expressed or implied, of the Army Research Laboratory or the U. S. Government.

Quotient Signal Estimation

D. Napoletani[1], C. A. Berenstein[2], P. Krishnaprasad[2], and D. C. Struppa[3]

[1] School of Computational Sciences
George Mason University
Fairfax, VA 22030
USA
`dnapolet@gmu.edu`
[2] Institute for Systems Research
University of Maryland
College Park, MD 20742
USA
`carlos@glue.umd.edu`, `krishna@glue.umd.edu`
[3] Department of Mathematical Sciences
George Mason University
Fairfax, VA 22030
USA
`dstruppa@gmu.edu`

Summary. In this paper, we propose a method for blind signal decomposition that does not require the independence or stationarity of the sources. We define suitable quotients of linear combinations of the images of the mixtures in a given frame and we show experimentally that such quotients can be used to recursively extract three sources from only two measurements. A general strategy to extract more than three sources from two measurements is proposed.

1 Introduction

Independent component analysis can recover signals that are linearly mixed with an unknown mixing matrix. All algorithms are essentially based on some local learning rule (see [L] and references therein, but also [QKS]). This procedure is effective, but it suffers from the need to assume that sources are independent and stationary. A different approach is taken in [CC], where sources are assumed to be independent and nonstationary and only time-delayed correlations of the observations are used to recover the mixing matrix. None of the previous methods can handle the case of mixtures of sources and their echoes, since clearly a source and its shifted versions are not independent.

In this paper, we suggest an algorithm that requires a different set of assumptions on the sources. This algorithm allows us to estimate more than two sources given at least two mixtures. More specifically, let $x_1[i] = a_1 s_1[i] + \cdots + a_N s_N[i]$, $x_2[i] = b_1 s_1[i] + \cdots + b_N s_N[i]$, $i = 1, \ldots, L$, be the two mixtures of N real-valued discrete sources $s_n[i]$, $n = 1, \ldots, N$, $i = 1, \ldots, L$, with L the length of the signals and $a_i, b_i \in \mathbb{R}$. Compute the expansion of x_1 and x_2 in some suitable frame dictionary $\mathcal{D} = \{g_1, \ldots, g_M\}$, $M > N$, for L-dimensional discrete signals, that is, compute the inner products $X_u[k] = \langle x_u, g_k \rangle$, $k = 1, \ldots, M$ and $u = 1, 2$.

Clearly, if we denote the expansion of s_n in $\mathcal{D}$ by S_n, we have

$$X_1 = a_1 S_1 + \cdots + a_N S_N, \qquad X_2 = b_1 S_1 + \cdots + b_N S_N.$$

Let R be a nonsingular 2×2 real-valued matrix that we call *exploratory matrix*, and consider the quotient

$$Q_R[k] = \frac{R(1, 1)X_1[k] + R(1, 2)X_2[k]}{R(2, 1)X_1[k] + R(2, 2)X_2[k]}, \qquad k = 1, \ldots, M.$$

In this paper, we use the collection of $Q_R[k]$ for some choice of R to find atoms in the dictionary where each source is dominant. This in turn reduces the search for the estimation of the sources to a type of projection pursuit. The reasoning behind our method is that if at k_0 there is only one source, say, S_{n_1}, then

$$Q_R[k_0] = \frac{R(1, 1)a_{n_1} S_{n_1}[k_0] + R(1, 2)b_{n_1} S_{n_1}[k_0]}{R(2, 1)a_{n_1} S_{n_1}[k_0] + R(2, 2)b_{n_1} S_{n_1}[k_0]} = \frac{R(1, 1)a_{n_1} + R(1, 2)b_{n_1}}{R(2, 1)a_{n_1} + R(2, 2)b_{n_1}},$$

that is, *the value of the quotient at k_0 is independent of the value of the measurements at that point, but it depends only on the coefficients a_{n_1}, b_{n_1} of S_{n_1}*. Ideally, if at each k only one source is dominant, the distribution of the values of Q_R is made up of several delta functions of different weight centered at 0 (due to the elements in the dictionary where there is no contribution from any signal) and at the value of the quotients

$$q_R(n) = \frac{R(1, 1)a_n + R(1, 2)b_n}{R(2, 1)a_n + R(2, 2)b_n},$$

$n = 1, \ldots, N$, that we can assume finite for a generic choice of R. To be able to discriminate the sources in the histogram of the value distribution of Q_R, it is therefore necessary that $\frac{b_n}{a_n} \neq \frac{b_m}{a_m}$ when $n \neq m$; we call $\frac{b_n}{a_n}$ the *slope* of the source s_n. In this ideal setting, the number of "dominant" peaks of the value distribution of the nonzero values of Q_R gives us an estimate of the number of sources and their positions will give the values of the $q_R(n)$ and therefore also the values of the slopes $\frac{b_n}{a_n}$. Unfortunately, in practice it is not possible to enforce a total separation of the sources in $\mathcal{D}$, therefore the value distribution of Q_R will at best give the position of the few most prominent peaks. Thus any effective algorithm based on the previous insights must be able to recursively reduce the influence of the sources on the shape of the histogram of Q_R. Even then, we need additional restrictions to avoid having degenerate situations in which ghost sources are detected. Assume, for example, that $S_m = pS_n$, p constant, $0 < m, n \le N$ in some region $\mathcal{M} \subset \mathcal{D}$ where the contribution of other signals is marginal. Then we have that on $\mathcal{M}$,

$$Q_R = \frac{R(1, 1)(a_n S_n + a_m S_m) + R(1, 2)(b_n S_n + b_m S_m)}{R(2, 1)(a_n S_n + a_m S_m) + R(2, 2)(b_n S_n + b_m S_m)}$$

$$= \frac{R(1, 1)(a_n + pa_m) + R(1, 2)(b_n + pb_m)}{R(2, 1)(a_n + pa_m) + R(2, 2)(b_n + pb_m)}.$$

The slope $\frac{b_n + pb_m}{a_n + pa_m}$ of a "source" that does not exist as a physical entity would be detected. Therefore, we must require that sources are linearly independent on most atoms of the dictionary. On the other hand, it is indeed possible that sources can be very similar in some cases; thus it is expected that there will always be limit cases that lead the algorithm into detecting ghost sources. Our own auditory system is not immune from illusions. We summarize now the previous discussion in a series of conditions that seem to be necessary for the histogram of the value distribution of Q_R to help in detecting the sources.

Condition (1). $\frac{b_i}{a_i} \neq \frac{b_j}{a_j}$ when $i \neq j$. We call $\frac{b_i}{a_i}$ the slope of the source s_i.

Condition (2). Each source is dominant on a subset of elements of $\mathcal{D}$.

Condition (3). Sources must be linearly independent for "most" values of $k = 1, \ldots, M$, i.e., we should not be able to find real numbers $(p_1, \ldots, p_n)$ such that $p_1 S_1[k] + \cdots + p_n S_n[k] = 0$.

These conditions are not fully rigorous; moreover, Conditions (2) and (3) are dependent on the choice of $\mathcal{D}$, which, as we will see, is changing in time in the algorithm we developed. Nevertheless such restrictions are the background against which we can assess the practical applicability of our method to specific problems. In the examples of the next section, something stronger than Condition (1) is enforced, namely, we select the mixing matrix in a set of matrices such that the angle between any two vectors (a_i, b_i) is larger than a given constant. We produce in this way matrices that are "nondegenerate" with respect to our method. Note that we do not require the sources to be independent or stationary, but rather we have geometrical separation conditions of the sources in the dictionary $\mathcal{D}$. In essence, two related data sets (X_1 and X_2) are projected onto a one dimensional space through the nonlinear function

$$Z = \frac{R(1, 1)X_1 + R(1, 2)X_2}{R(2, 1)X_1 + R(2, 2)X_2}.$$

Therefore, we can view the underlining method as a type of nonlinear projection pursuit in which the choice of the exploratory matrix determines the specific nonlinear projection of interest (see [H] for an extensive treatment of projection pursuit). In a way, we can say that our approach relaxes the requirements on the sources while it imposes more stringent conditions on the mixing matrices.

The second section of this paper introduces the general strategy of the "quotient projection" algorithm. In Section 3, we perform experimental separations for the case of two mixtures and three sources. Moreover, we show that, for the case of speech sources, a more sophisticated technique is needed if we want to extract four sources from two measurements. The strategy of such generalized algorithm is outlined.

The possibility of identifying and separating sources through the use of the simple quotient $\frac{X_1}{X_2}$ in the time frequency domain was already underlined by Rickard and collaborators in [RD, RBR, RY, BR], but see also our preliminary unpublished work [NBK]. In this paper, on the other hand, we stress the possibility of choosing *several* exploratory matrices to build a robust algorithm that can be implemented in *general* dictionaries. Especially the results in Section 3 show that any one single choice of exploratory matrix is unable to properly identify all sources for most mixing matrices.

It should be noted that a major problem for speech (audio) signals is to find the correct model of their mixing. Our assumption here (implicit in the way in which the mathematical model is constructed) corresponds to instantaneous mixing; while this choice is common in testing the basic performance of blind signal separation algorithms, it is very restrictive for real-world signals, where one needs to take into account echoes and asynchrony of measurements.

2 Quotients projections

We directly give a description of the basic steps of our algorithm (based on the insights of the previous section). We then explain the heuristics behind it.

(a) Given x_1 and x_2, choose a dictionary $\mathcal{D}$ and an exploratory matrix R.
(b) Compute X_1 and X_2, compute the quotient Q_R, and find the first N "significant" maxima q_n, $n = 1, \ldots, N$, of the histogram of the value distribution of Q_R.
(c) Set the initial estimate of sources $\bar{s}_n$ as $\bar{s}_n[i] = 0$, $i = 1, \ldots, L$, $n = 1, \ldots, N$.
(d) Choose positive numbers ϵ_n and let $\mathcal{T}_n$ be the collections of all g_k in $\mathcal{D}$ such that $|Q_R[k] - q_n| < \epsilon_n$.
(e) Let $Z_{n,u}[k] = X_u[k]$, $u = 1, 2$, and $n = 1, \ldots, N$ if $g_k \in \mathcal{T}$ and $Z_{n,u}[k] = 0$ otherwise. Compute the inverses $z_{n,u}$ of $Z_{n,u}$, $u = 1, 2$, $n = 1, \ldots, N$.
(f) Let $\bar{s}_n = \bar{s}_n + z_{n,1}$, $n = 1, \ldots, N$, and set $x_u = x_u - \sum z_{n,u}$, $u = 1, 2$. Go to step (a).

The first thing to note is that (a)–(f), as it is, does not have a termination rule. In the experiments performed in the next section, we iterated the core loop (a)–(f) a fixed large number of times. Note also that the selection of $\mathcal{D}$ and R in (a) does not need to be the same at every iteration. Actually, we will see in the examples of the following section that there is an advantage in selecting several dictionaries to find all atoms that contribute to a given source. The same reasoning can be applied to the choice of R: We will give evidence in the next section that the most general strategy based on (a)–(f) uses several choices of R. The algorithm (a)–(f) first identifies the dominant sources in step (b), with the inspection of the histogram of Q_R, and then it uses the localization of the peaks associated with these sources in steps (d)–(e) to "extract" all of the structure belonging to such dominant sources as visible in the dictionary $\mathcal{D}$. The whole process is repeated with another dictionary in a fashion similar to matching pursuit algorithms [M]. Step (b) is ambiguous as we do not explain how to find the "significant" sources; the fact is that the number of sources that can be extracted is limited in any case when the sources are speech signals (our main case of study). In our implementation, we assume we know a priori the maximum number N_0 of sources, and we build a robust estimator of the first significant $N < N_0$ maxima of the histogram of the value distribution of Q_R. The specific implementation of step (b) of our procedure is the following.

(b′) Compute Q_R. Build a best estimation of the value distribution of Q_R between Q_5 and Q_{95}, respectively, the 5 and 95 percentiles (to avoid the selection of large domains due to extreme outliers). Choose the width of the bins of the histogram to be $\frac{Q_{95}-Q_5}{500}$, this bin size is fine enough provided R is close to a singular matrix, so that the true quotients are effectively close to each other. Let H_R be the resulting histogram. We then use smoothing splines to find the best approximation of H_R that has no more than N_0 maxima. Let q_n, $n = 1, \ldots, N < N_0$, be the ordinate of such maxima, let (l_n, r_n) be the inflection points left and right of q_n. Let $m_n = \min(|H_R(q_n) - H_R(l_n)|, |H_R(q_n) - H_R(r_n)|)$, and let $M = \max_n m_n$; we consider significant only the maxima such that the $m_n > 0.05M$. Moreover, we set $\epsilon_n = \frac{1}{2} \min(|q_n - l_n|, |q_n - r_n|)$.

It is immediately evident that there are several parameters chosen in (b′) in a somewhat arbitrary way; nevertheless (b′) as stated seems to work for large classes of mixing matrices and for most speech sources we tested. Further experimental and theoretical work would be necessary to establish if (b′) is indeed the most efficient implementation of (b). Here we say only that often the true quotient $q_R(n)$ will not be detected as a maximum of H_R, but it will appear as sizeable asymmetrical "enlargement" of one of the other peaks (see Figure 6 in the next section). These enlargements are associated to inflection points of the smoothened histogram, so our choice of ϵ_n as half the distance of the computed quotients q_n from the nearest inflection point prevents the most common cause of clustering two separate sources in a single reconstruction.

Remark. We restrict our attention to the case in which all (a_n, b_n) are in the positive quadrant, as this case corresponds to the most relevant applications in speech processing, where the coefficients of the mixing matrix are positive attenuation coefficients of the energy intensity. Given the previous restriction, any fixed choice of R such that r_1 and r_2 are properly contained in the positive quadrant does assure that there is a lower bound on $\langle v_i, r_j \rangle$, $j = 1, 2$, for any possible v_i in the positive quadrant. Note that the choice of

$$R = \begin{bmatrix} 1 & 0 \\ 0 & 1 \end{bmatrix},$$

which is the simple quotient $\frac{X_1}{X_2}$, would reduce the ability of identifying any source s_n such that $(a_n, b_n) \approx (1, 0)$ or $(a_n, b_n) \approx (0, 1)$ since in the first case s_n would have marginal contribution in the denominator, and in the second case the signal would be marginal in the numerator and there would be no significant peak associated with $q_R(n)$ in the histogram of the value distribution of Q_R.

Remark. In [BR], the authors offer an interesting stochastic model for mixing of speech signals that in the case of the quotient $\frac{X_1}{X_2}$ provides an optimal estimation of the position of the peaks. It is conceivable that similar ideas could be used in our case, instead of the heuristic in (b′), to find the position of the peaks, the identification of ϵ_n in (d) above, and possibly to select the optimal exploratory matrix as well.

3 Experiments and further developments

In this section, we treat two distinct cases: the case of three sources of interest and two measurements, and the case in which there are four sources and two measurements. In the second case, we give evidence that no single choice of exploratory matrix is enough to find all four quotients.

In the following two experiments, we apply the algorithm to speech signals from the TIMIT database. Because of this choice, the dictionary in (a) is selected to be a cosine packets or wavelet packets basis (see [M]) of the sum of the current x_1 and x_2. These types of bases are known to be effective in approximating speech signals, and therefore they increase the chance of having Condition (2) satisfied. Moreover, to increase the robustness of the iterative structure extraction, we attenuate the coefficients selected in (e) and we alternate between wavelet packets bases and cosine packets bases in (a). We explicitly write these modifications of steps (a) and (e):

(a′) Given x_1 and x_2, compute $x = x_1 + x_2$. Choose a "random" cosine or wavelet packets basis $\mathcal{D}$ for x. Choose an exploratory matrix R.

(e′) If the basis used is a cosine packets basis, set $\alpha = 0.4$; otherwise, if it is a wavelet packets basis, set $\alpha = 0.1$. Let $Z_{n,u}[k] = \alpha X_u[k]$, $u = 1, 2$, and $n = 1, \ldots, N$ if $g_k \in \mathcal{T}$ and $Z_{n,u}[k] = 0$ otherwise. Compute the inverses $z_{n,u}$ of $Z_{n,u}$, $u = 1, 2$, $n = 1, \ldots, N$.

Remark. The values of α for cosine and wavelets packets were heuristically adjusted to obtain good reconstruction when the algorithm alternates one iteration using cosine packets and one using wavelets packets. With regard to the "random" choice of basis in (a′), in this paper we simply selected a small set of basis trees and alternated among them uniformly to generate our cosines and wavelets packets bases. The issue of basis selection deserves certainly a more detailed study, and in a subsequent paper we will compare the performance of several different ways to choose dictionaries in (a′). We are now ready to write the basic implementation of our algorithm:

(A1) Apply (a) and (b′) once with the windowed Fourier frame (see [S]) as the choice of dictionary $\mathcal{D}$.

(A2) Apply (a′), (c)–(d), (e′), and (f) a fixed number ($T = 60$) of times, alternating in (a′) a choice of cosine packets and wavelet packets basis.

We first expand in the windowed Fourier frame (rather than a basis) to assure the generation of a relatively smooth histogram determined by *local* time frequency structure. Note that the identification of the dominant regions in the windowed Fourier frame are used only to "track" the positions of the quotients $q_R(n)$: The sources do not need to be totally separated in any given basis. Rather, we use several bases in (A2) to recursively separate the structure belonging to different sources.

Example 1. In our first set of experiments, we ask for the estimation of three unknown speech sources (of a duration of about one second) given two measurements, a situation that, in the setting of this paper, can be described by the model

$$x_1 = a_1 s_1 + a_2 s_2 + a_3 s_3, \qquad x_2 = b_1 s_1 + b_2 s_2 + b_3 s_3.$$

We apply the algorithm to 30 instances of the model with a_u, b_u, $u = 1, 2, 3$, chosen as coordinates of unit vectors in the positive quadrant such that the angle between any two of them is at least $\pi/16$. The specific way we generated such mixing matrices is the following: We repeatedly compute three angles θ_n, $n = 1, 2, 3$, chosen uniformly in $[0, \frac{\pi}{2}]$ until their minimum angular distance is bigger than $\frac{\pi}{16}$. We then set $a_n = \sin(\theta_n)$ and $b_n = \cos(\theta_n)$. The choice of the exploratory matrix is taken as

$$R = \begin{bmatrix} \sin\left(\dfrac{\pi}{4}\right) & \cos\left(\dfrac{\pi}{4}\right) \\[2ex] \sin\left(\dfrac{\pi}{4} + \delta\right) & \cos\left(\dfrac{\pi}{4} + \delta\right) \end{bmatrix},$$

for some $\delta \ll \frac{\pi}{16}$. The reason we take both numerator and denominator close to $(\sin(\frac{\pi}{4}), \cos(\frac{\pi}{4}))$ is that for this choice we have the largest possible lower bound on the reduction of the norm of the projection of any vector in the positive quadrant, so avoid, as much as we can, to unwittingly reduce the contribution of one source to the overall value distribution of Q_R.

Figure 1 shows the generic smoothened histogram generated by (b'). The attenuation coefficient α in (e') prevents a full recovery of the energy of the sources, but what we are really interested in is of course the overall shape of the reconstruction; thus in Figures 2, 3, and 4 we show from top to bottom the original sources and the reconstructions (for one representative choice of mixing matrix) scaled to have norm 1. In Figure 5, we show the two original mixtures. Since the comparison between sources and reconstructions is meaningful only modulo a rescaling of them, we define the scaled signal-to-noise ratio (scaled SNR) to be the usual SNR (measured in decibels), where signal and reconstruction are scaled to have norm 1.

We can summarize our results by noting that in all cases except one, we find three maxima, and therefore we obtained reconstructions of all sources s_1, s_2, and s_3 96% of the time. The mean scaled SNR was 7.6db for reconstructions of s_1, 6.4db for reconstructions of s_2, and 6.3db for reconstructions of s_3. The perceptual quality was very good for most reconstructions.

Incidentally, the same set of 30 mixtures, if analyzed with the exploratory matrix

$$R = \begin{bmatrix} 1 & 0 \\ 0 & 1 \end{bmatrix},$$

leads to lower scaled SNR and a lower proportion of reconstructions: We reconstruct exactly one source 20% of the time, exactly two sources 43% of the time, and just 37% of the time we recover exactly three sources. On the other hand, the mean scaled SNR for the cases in which we do have reconstruction is only slightly worse than for the exploratory matrix of Example 1. We have a scaled SNR of 7.6db for the nonzeros reconstructions of s_1, 5.8db for the nonzeros reconstructions of s_2, and 6.2db for the nonzeros reconstructions of s_3.

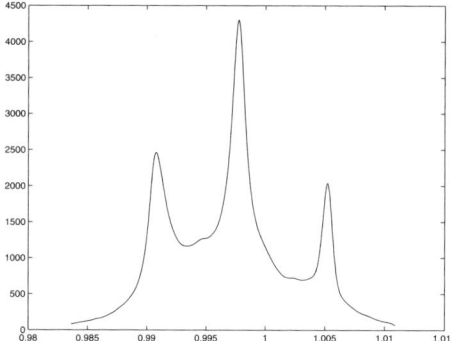

Fig. 1. Generic smoothened histogram from step (b′) as described in the text with three sources.

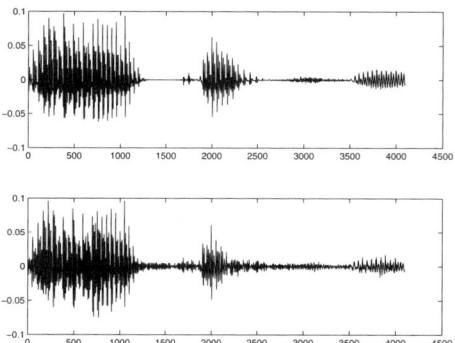

Fig. 2. From top to bottom, signal s_1 and one instance of reconstruction $\bar{s}_1$, both scaled to have norm 1.

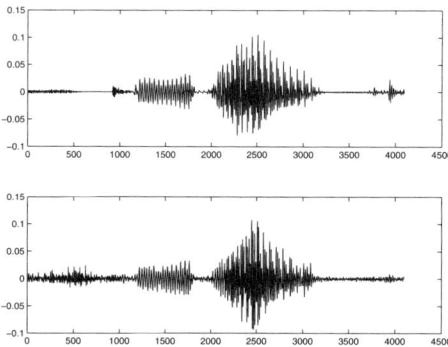

Fig. 3. From top to bottom, signal s_2 and one instance of reconstruction $\bar{s}_2$, both scaled to have norm 1.

Example 2. The generalization to the recovery of more than three sources in the setting of algorithm (A1)–(A2) is problematic. Assume, for example, that we have four specific speech sources s_i, $i = 1, \ldots, 4$ (of a duration of about one second), and

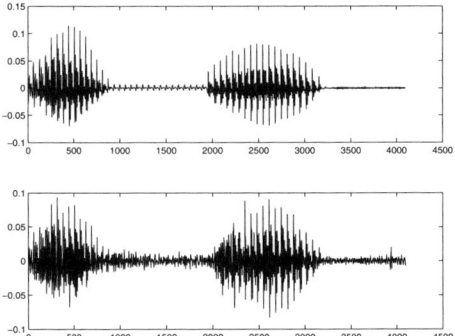

Fig. 4. From top to bottom, signal s_3 and one instance of reconstruction $\bar{s}_3$, both scaled to have norm 1.

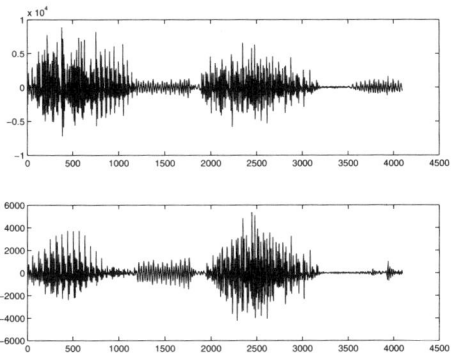

Fig. 5. Original measurements.

two mixtures,

$$x_1 = a_1 s_1 + a_2 s_2 + a_3 s_3 + a_4 s_4, \qquad x_2 = b_1 s_1 + b_2 s_2 + b_3 s_3 + b_4 s_4.$$

All other details are as in Example 1, modulo the increase in the number of sources. The result of applying algorithm (A1)–(A2) to this setting is somewhat disappointing. We reconstruct exactly one source 20% of the time, exactly two sources 54% of the time, exactly three sources 23% of the time, and just 3% of the time we recover exactly four sources. The mean scaled SNR is 4db for reconstructions of s_1, 4.2db for reconstructions of s_2, 3.7db for reconstructions of s_3, and 4.2db for reconstructions of s_4. In Figure 6, we show a typical histogram generated by (b'). We see clearly that the procedure fails to detect one of the quotients, which appears as an inflection point at the right of largest maxima.

We may think that a diminishing quality of the reconstructions is unavoidable as the number of sources increases—after all, we are dealing with more and more underdetermined systems of vector equations. On the other hand, we do believe that at least the attempt of reconstructing four speech sources from two measurements is

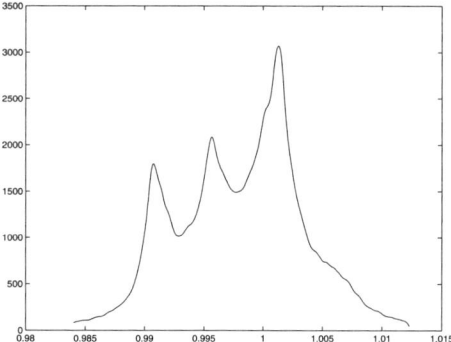

Fig. 6. Generic smoothened histogram from step (b′) as described in the text with four sources, *but only three detected maxima.*

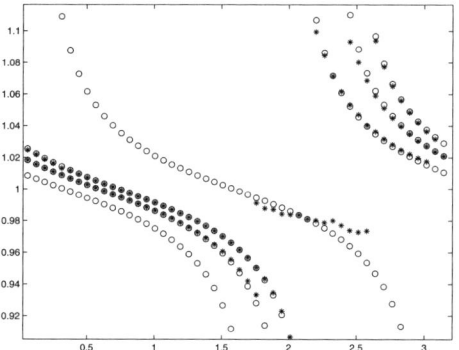

Fig. 7. Curves of true ("o" curves) and computed ("∗" curves) quotients as the parameter β goes from 0 to π.

within the reach of our method. In Figure 7, we show the position of true ("o" curves) and computed ("∗" curves) quotients as we change a particular unidimensional family of exploratory matrices for a specific fixed choice of mixing matrix. We assume that the angular distance between coefficients of the sources is larger than $\frac{\pi}{16}$ (as we enforced in Examples 1 and 2). The family of exploratory matrices has the form

$$R_\beta = \begin{bmatrix} \sin\left(\beta + \dfrac{\pi}{100}\right) & \cos\left(\beta + \frac{\pi}{100}\right) \\ \sin(\beta) & \cos(\beta) \end{bmatrix},$$

where β goes from 0 to π. That is, R_β is made of two vectors that are much closer to each other than any two vectors (a_n, b_n).

Note that each of the true quotient curves γ_n is defined exactly modulo π and that it has a very regular monotone shape, with divergence corresponding to the value of β that makes $(\sin(\beta), \cos(\beta))$ perpendicular to (a_n, b_n). Also note that there is at least one value of R_β for which the computed quotient curve $\bar{\gamma}_n$ is locally well approximating the corresponding γ_n. While *there is no value of R_β for which all curves are locally approximated*. This observation is the key of a refinement of algorithm (A1)–(A2) that we sketch in (B1)–(B4).

(B1) Apply (a) once with the windowed Fourier frame as choice of dictionary $\mathcal{D}$.
(B2) Apply (b′) for the entire family of exploratory matrices R_β. For each computed significant quotient q, let $\bar{\gamma}_n(\beta, q) = H_R(q)$ be the intensity of the histogram at the quotient itself.
(B3) For each curve $\bar{\gamma}_n$, find the value of $\beta = \beta_n$ that makes $\bar{\gamma}_n(\beta_n, q)$ maximum.
(B4) For each n, choose $R = R_{\beta_n}$ and apply (a′), (c)–(d), (e′), and (f) a fixed number ($T = 60$) *only to the estimation of source $\bar{s}_n$*. Alternate in (a′) a selection of cosine and wavelet packets best basis.

Preliminary results show that (B1)–(B4) is effective in extracting four speech sources from two measurements most of the time; in a subsequent paper, we will return to a complete implementation and analysis of this "moving" quotient algorithm.

Data files of the sources, mixtures, and reconstructions computed in Examples 1 and 2 are available upon request for direct evaluation of their perceptual quality.

Acknowledgments This research was supported in part by the Office of Naval Research under ODDR&E MURI97 Program Grant N000149710501EE to the Center for Auditory and Acoustics Research and by the Army Research Office under ODDR&E MURI01 Program Grant DAAD19-01-1-0465 to the Center for Communicating Networked Control Systems (through Boston University). The second author was also supported in part by National Science Foundation grant NSF-DMS-0070044. We would like to thank Shihab Shamma and Marco Panza for valuable and stimulating discussions. We are also grateful to the referee for constructive and valuable suggestions.

References

[BR] R. Balan and J. Rosca, Statistical properties of STFT ratios for two channel systems and applications to BSS, in *Proceedings of ICA 2000*, Helsinki, 2000.

[CC] S. Choi and A. Cichoki, Blind separation of nonstationary sources in noisy mixtures, *Electron. Lett.*, **36**-9 (2000), 848–849.

[D] D. Donoho, Sparse components of images and optimal atomic decompositions, *Constr. Approx.*, **17**-3 (2001), 353–382.

[GG] I. J. Good and R. A. Gaskins, Nonparametric roughness penalties for probability densities, *Biometrika*, **58**-2 (1971), 255–277.

[H] P. J. Huber, Projection pursuit, with discussion, *Ann. Statist.*, **13** (1985), 435–525.

[HR] J. H. van Hateren and D. L. Ruderman, Independent component analysis of natural image sequences yields spatiotemporal filters similar to simple cells in primary visual cortex, *Proc. Roy. Soc. London Ser. B*, **265** (1998), 359–366.

[L] T.-W. Lee, *Independent Component Analysis: Theory and Applications*, Kluwer, Boston, 1998.

[M] S. Mallat, *A Wavelet Tour of Signal Processing*, Academic Press, New York, 1998.

[NBK]   D. Napoletani, C. A. Berenstein, and P. S. Krishnaprasad, *Quotient Signal Decomposition and Order Estimation*, Technical Report 2002-47, Institute for Systems Research, University of Maryland, College Park, MD, 2002; available online from techreports.isr.umd.edu/ARCHIVE/.

[OF]   B. A. Olshausen and D. J. Field, Sparse coding with an overcomplete basis set: A strategy employed by V1?, *Vision Res.*, **37** (1997), 3311–3325.

[QKS]   Y. Qi, P. S. Krishnaprasad, and S. Shamma, The subband-based independent component analysis, in *Proceedings of ICA 2000*, Helsinki, 2000.

[RD]   S. Rickard and F. Dietrich, DOA estimation of many W-disjoint orthogonal sources from two mixtures using duet, in *Proceedings of the 10th IEEE Workshop on Statistical Signal and Array Processing*, IEEE, New York, 2000, 311–314.

[RBR]   S. Rickard, R. Balan, and J. Rosca, Real-time time-frequency based blind source separation, in *Proceedings of ICA 2001*, San Diego, 2001.

[RY]   S. Rickard and O. Yilmaz, On the approximate W-disjoint orthogonality of speech, in *Proceedings of the 2002 IEEE International Conference on Acoustics, Speech and Signal Processing*, Vol. 1, IEEE, New York, 2002, 529–532.

[S]   T. Strohmer, Numerical algorithms for discrete Gabor expansions, in H. G. Feichtinger and T. Strohmer, eds., *Gabor Analysis and Algorithms: Theory and Applications*, Birkhäuser, Boston, 1998.